High-Speed Photonic Devices

Series in Optics and Optoelectronics

Series Editors: **R G W Brown**, University of Nottingham, UK
E R Pike, Kings College, London, UK

Other titles in the series

Series in Optics and Optoelectronics

High-Speed
Photonic Devices

Nadir Dagli
University of California
Santa Barbara, USA

CRC Press
Taylor & Francis Group
Boca Raton London New York

CRC Press is an imprint of the
Taylor & Francis Group, an **informa** business

CRC Press
Taylor & Francis Group
6000 Broken Sound Parkway NW, Suite 300
Boca Raton, FL 33487-2742

First issued in paperback 2019

© 2007 by Taylor & Francis Group, LLC
CRC Press is an imprint of Taylor & Francis Group, an Informa business

No claim to original U.S. Government works

ISBN-13: 978-0-7503-0889-2 (hbk)
ISBN-13: 978-0-367-39027-3 (pbk)

Library of Congress Cataloging-in-Publication Data

Dagli, N. (Nadir)
   High-speed photonic devices / author : N. Dagli.
      p. cm.
   Includes bibliographical references and index.
   ISBN-13: 978-0-7503-0889-2 (alk. paper)
   ISBN-10: 0-7503-0889-3 (alk. paper)
   1. Optoelectronic devices. 2. Very high speed integrated circuits. I. Title.

TK8304.D33 2006
621.381'045--dc22
                                               2006011669

Visit the Taylor & Francis Web site at
http://www.taylorandfrancis.com

and the CRC Press Web site at
http://www.crcpress.com

Editor

Nadir Dagli received his Ph.D. degree in electrical engineering from Massachusetts Institute of Technology in 1987.

After graduation he joined the electrical and computer engineering department at University of California at Santa Barbara, where he is currently a professor. He has done pioneering work on finite difference beam propagation method; ultra fast substrate removed compound semiconductor electro-optic modulators, slow wave electrodes for efficient optical modulators, electron waveguides and couplers, total internal reflection mirrors for compact photonic integrated circuits and compact ring resonators integrated with semiconductor optical amplifiers. His current interests are design, fabrication and modeling of guided-wave components for optical integrated circuits, ultrafast electro-optic modulators, WDM components, and photonic nanostructures.

Dagli was awarded NATO science and IBM predoctoral fellowships during his graduate studies. He received the 1990 UCSB Alumni Distinguished Teaching Award and 1990 UC Regents Junior Faculty Fellowship. He served as a technical program committee member, program and conference chair, and advisory board member of many international conferences including the Integrated Photonics Research Conference, SPIE's Photonics West, International Topical Meeting on Microwave Photonics, CLEO, CLEO/Pacific Rim, IEEE Lasers and Electro Optics Society Annual Meeting. He was the associate editor of IEEE Photonics Technology Letters from 1997 to 2000 and the editor-in-chief of IEEE Photonics Technology Letters from 2000 to 2005. He also served as an elected member of the IEEE-LEOS board of governors from 2003 to 2005. He is a fellow of IEEE.

Contributors

Nadir Dagli
Electrical and Computer Engineering
  Department
University of California at Santa
  Barbara
Santa Barbara, CA

P. Fay
Department of Electrical
  Engineering
University of Notre Dame
Notre Dame, IN

Harold R. Fetterman
Department of Electrical
  Engineering
University of California,
  Los Angeles
Los Angeles, CA

Osamu Mitomi
Ubiquitous Network Group
NGK Insulators, LTD
Nagoya, Japan

Taiichi Otsuji
Center for Microelectronic Systems,
  Faculty of Computer Science and
  Systems Engineering
Kyushu Institute of Technology
Fukuoka, Japan

Masatoshi Saruwatari
National Defense Academy
Yokosuka-shi, Kanagawa Japan

William H. Steier
Department of Electrical
  Engineering
University of Southern California
Los Angeles, CA

Koichi Wakita
Department of Electronic Engineering
Chubu University
Kasugai, Aichi, Japan

Table of contents

1 Introduction

Nadir Dagli

Ever since the Internet was opened to public use in 1993, the amount of information transmitted over communication networks has increased drastically. Most of the traffic in communication networks today is due to data transmission rather than voice transmission, and the amount of data transmitted continues to increase rapidly. For such high-volume applications, fiber is the most natural transmission medium. Standard single-mode fiber offers more than 25 terahertz bandwidth. The availability of such high bandwidth enables very high-bit-rate transmission. Fiber optic networks operating at 40 Gbit/second are being installed all around the globe. The technology that provides such high-bit-rate transmission is crucial and is being developed in many different research laboratories around the world. This technology requires high-speed transmitters, high-speed receivers, high-speed electronics, and high-speed all-optical techniques. It is obvious that these topics require very diverse backgrounds, and it is best to rely on the expertise of individuals who are leaders in their fields. This book presents this technology by combining several chapters written by the experts working on different aspects of high-speed photonic devices.

High-speed transmission starts at the source end. In principle, a high-speed transmitter can be obtained by directly modulating a semiconductor laser diode. Although semiconductor lasers with small signal bandwidths on the order of 40 GHz have been demonstrated, direct modulation is typically not used for transmissions over 2.5 Gbit/second. This is mainly due to chirping of the laser output under direct modulation. As the current supplied to the laser diode is modulated, the number of carriers in the device changes. This changes the index of the material and the laser output frequency. This undesired frequency modulation increases the bandwidth of the transmitted signal and severely limits the transmission distances. Therefore, for high-speed transmission systems, external modulators are used, and the laser output is kept unchanged. High-speed external modulation can be achieved using different technologies. These technologies are the topics of Chapter 2 through Chapter 6.

Chapter 2 describes electroabsorption (EA) modulators and EA modulators integrated with distributed feedback (DFB) lasers. The EA modulator is a very compact semiconductor device. The absorption through this device depends on the voltage applied to the device. This absorption modulation can be done very efficiently; even a few volts can be sufficient. EA modulators typically use specially designed quantum-well absorption regions. Therefore, proper material design and growth is crucial. They work efficiently over a limited range of wavelengths and demonstrate sensitivity to temperature changes and high optical input powers. They also tend to have high optical-insertion loss. EA modulators are compact (in most cases less than 200 μm long) and can be integrated with DFB lasers. This results in a very compact source that can be modulated very efficiently. Such devices at bit rates of 10 Gbit/second are commercially available from several different vendors. EA modulators also have chirp, which varies during modulation. However, their chirp can be kept small and negative, which actually may help in data transmission to a certain degree.

The third chapter describes LiNbO$_3$ modulators. LiNbO$_3$ is a ferroelectric material that possesses an electro-optic coefficient. The index of refraction of the material changes in response to an applied electric field without changing the optical loss. Therefore, it is possible to make interferometric or Mach-Zehnder–type modulators in this material. Such modulators can have zero chirp and very low insertion loss. Furthermore, LiNbO$_3$ has a very large electro-optic coefficient. However, even in this case an efficient modulator requires a traveling-wave electrode. In such designs the modulator electrode is designed as a transmission line, and the modulating voltage and the modulated optical wave travel in the same direction and interact along the length of the electrode. If the optical and electrical velocities are matched, this interaction is most efficient and the small phase changes can be integrated along the length of the electrode. Typical electrode lengths for LiNbO$_3$ modulators are approximately 3 to 5 centimeters. This approach helps to reduce the drive voltage to a few volts but results in large devices. One fundamental difficulty of this material is its dielectric constant dispersion. The dielectric constant at microwave and millimeter wave frequencies is very high, but it decreases significantly at optical frequencies. As a result, the electrical signal travels slower than the optical signal, and velocity matching requires increasing the electrical signal velocity. Therefore, electrode capacitance should be reduced, which in turn requires reducing the electric field for a given voltage. The large electro-optic coefficient of LiNbO$_3$ partially offsets these difficulties. Such devices are commercially available from several vendors.

The fourth chapter covers III-V compound semiconductor modulators. These materials also possess an electro-optic coefficient unlike the most common semiconductor, silicon, which does not. Their electro-optic coefficient is smaller than that of LiNbO$_3$, but they have high indices of refraction and very low index dispersion between microwave and optical frequencies. Since electro-optic index change is proportional to the cubic power of the index of refraction, the high index of refraction makes up for the low electro-optic coefficient to a certain extent. Furthermore, the low-index dispersion makes the electrical signal travel faster than the optical signal. As a result, traveling wave designs with slow wave electrodes become possible, and large capacitive loading can be used. The high index of refraction also makes tight optical confinement possible. When these two effects are combined, it becomes possible to have tightly confined optical modes overlapping very well with very high modulating fields. Therefore even bulk III-V compound semiconductor modulators with better voltage-length products than

LiNbO$_3$ are possible. Furthermore, the electro-optic coefficient of these materials can be improved using quadratic effects that are related to electroabsorption. These devices also have the potential of integration with other III-V compound semi-conductor devices such as lasers and detectors. However they have larger insertion loss mainly due to small-mode size-related coupling loss. Mode transformers can solve this problem, but at increased cost.

Chapter 5 discusses polymer modulators. Polymers offer the potential of both a high electro-optic coefficient and a low index of refraction. Their high electro-optic coefficient has the potential to reduce the drive voltage. Their low index of refraction should help to increase the modulation bandwidth since velocity matching becomes easier and microwave electrode loss tends to be lower on lower-index material. Synthesizing high electro-optic coefficient polymers has been hampered due to reliability issues for quite some time. However, recent advances have solved this problem to a certain extent, and good-quality polymers with reasonably high electro-optic coefficients are available. The design of modulators and other guided wave components is described in detail in this chapter.

The critical components at the end of a transmission system are the photodetectors and photoreceivers, which are described in Chapter 6. These components convert optical information into electrical information while minimizing noise and distortion. There are many different types of photodetectors used for different types of applications. A pho-todetector is followed by an electronic circuitry for amplification and signal processing. The receiver design requires the selection of appropriate technology, devices, and archi-tecture. Recently, different modulation formats have been introduced to improve the spectral efficiency of optical transmission. Special receivers are needed for such trans-mission systems.

Chapter 7 covers integrated circuit technologies for fiber optic systems. High-speed electronic integrated circuits are essential for the realization of fiber optic systems. Both receiver and transmitter ends require high-speed electronics not only for modulator drivers and photoreceivers, but also for clock recovery and decision making. Currently, electronic integrated circuits operating at 40 Gbits/second are available in different technologies including Si-, Si-Ge-, GaAs- and InP-based mate-rials. Their fundamental limitations, advantages, and architectures are the topic of this chapter. This chapter also discusses the limiting factors for electronic integrated circuits for 100-Gbit/s and faster operation.

The last chapter is on all-optical high-speed technologies. As demand for infor-mation increases, electronic limitations could be the bottleneck for high-speed fiber optic networks. This limitation could be overcome using all-optical techniques not only in transmission but also in processing information at the network nodes. Optical signal processing has much wider bandwidth than electronic processing and has the potential of providing extremely high-speed transmission systems. In addition to speed, transparency for modulation format is another very important property of networks using all-optical techniques. Ultrashort and ultrahigh-speed optical pulse generation, modulation, multiplexing, and demultiplexing are technologies for high-speed transmission. These technologies coupled with other all-optical technologies such as optical timing extraction, optical signal monitoring, all-optical repeating, and all-optical regenerating could enable ultrafast networks.

After reading the book, it will become obvious that there is strong potential for significant further improvement in all the topics considered. Therefore, all high-speed photonic components represent active and exciting research areas. Although it is hard to predict which particular technology will prevail in the future, it is expected that these research efforts will result in low-cost, high-performance fiber optic networks covering the entire globe. Such improved and easily accessible communications capability will improve the quality of life for everyone.

Finally, the editor takes this opportunity to thank all the contributors to this book for taking their time to write very informative chapters about their work.

2 Electroabsorption (EA) Modulators and EA Modulators Monolithically Integrated with Distributed-Feedback Lasers

Koichi Wakita

CONTENTS

ABSTRACT

This chapter describes the recent research and development of high-speed, low-driving-voltage EA modulators and their integrated distributed-feedback lasers for high-bit-rate and long-haul optical fiber transmission systems. Low insertion loss and low chirping as well as polarization insensitivity are also discussed. Increased saturation intensity, high-input optical power allowance, and wide-wavelength operation for wavelength-division multiplexing are described. Particular attention has been paid to multiple quantum well (MQW) modulators operating at long wavelengths, taking into account the losses and dispersion in silica fibers occurring at around 1.3 and 1.55 μm. The future prospects for optical modulators and their monolithically integrated light sources for ultrahigh-bit-rate OTDM are also discussed.

2.1 INTRODUCTION

The research and development of optical fiber communication systems, satellite communication systems, and radio transmission systems based on large-scale integrated circuit technologies has opened the door to a new age in the application of Internet and multimedia communications. In the early 21st century, telecommunications and computers will be combined into information network systems to satisfy the demands of the enormous increases in information and communication required by society and industry. Semiconductor modulators mainly used as electroabsorption (EA) modulators are one of the important devices playing a role as key components of this new age. EA modulators integrated with DFB lasers have been developed and applied to commercial uses for metropolitan high-bit-rate and long-haul optical transmission systems. Moreover, in optical-time-division-multiplexed (OTDM) systems based on high-speed optical switching (over 100 Gbit/sec), EA modulators are indispensable, and much interest has been focused on them.

In this chapter, the progress of the semiconductor optical modulators throughout eight recent years [1] will be reviewed from the viewpoints of high-speed and low-driving-voltage operation. New directions of modulator research such as low-chirp operation, high-input optical power allowance, and wide wavelength operation for wavelength-division multiplexing (WDM) as well as low insertion loss will also be discussed.

2.1.1 FRANZ-KELDYSH EFFECT

The research and development of EA modulators has a long history. In 1958, Franz and Keldysh [2,3] independently reported groundbreaking theoretical studies of the effect of an electric field on the absorption edge of a semiconductor. They predicted that, in the presence of an electric field, an absorption would occur at photon energies lower than the forbidden energy gap (Franz-Keldysh effect), as shown in Figure 2.1.

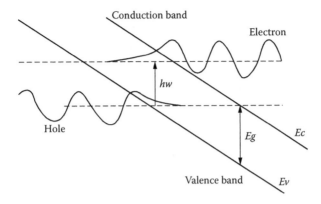

FIGURE 2.1 Energy band diagram of a semiconductor under high electric field.

However, the bulk structure used at that time had a large operating voltage and large insertion loss. The limitations of directly modulated laser diodes have caused us to take a new look at this mechanism. In high-bit-rate (over a few Gbit/s), long-haul optical transmission systems, the frequency chirping associated with high-speed direct modulation of semiconductor laser diodes is a serious problem that limits transmission length and modulation speed. Advances in double heterostructure growth by liquid-phase-epitaxy and waveguide theory brought about renewed interest in applying the effect for light modulation, and there were many attempts at devising the effect for light modulators [4–8]. Several volts operation for an on/off ratio of more than 10 dB has been achieved. Optical fiber transmission using EA modulators that were monolithically integrated with DFB laser diodes was also demonstrated [9–11]. However, driving voltages are large compared with those of MQW modulators.

2.1.2 Quantum-Confined Stark Effect

The multiple quantum well (MQW) structure exhibits strong excitonic effects that modify the fundamental absorption edge of materials. The exciton effect results from the Coulomb interaction between the electron-hole pair in the crystal, and it manifests itself as an increased steepness in the absorption coefficient. These effects are also present in covalent semiconductors, but they can be observed only at very low temperatures due to the low excitonic binding energy (typically 3 meV for Ge and GaAs). In MQW structures, the binding energy can be as large as 10 meV and the excitonic structure can be observed at room temperature. When an electric field is applied perpendicularly to such structures, the absorption coefficient near the band edge decreases significantly, as shown in Figure 2.2. This is similar to the Franz-Keldysh effect; however, because it is associated with quantum well (QW) structures, it is called the quantum-confined Stark effect (QCSE) [12]. The difference between the two is that, due to the two-dimensional confinement, excitonic peaks and their energy-shifts are observed for MQW structures, while no such effects are observed for the Franz-Keldysh effect. The energy shift is approximately proportional to the

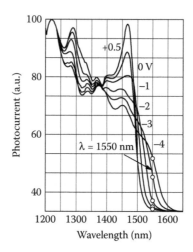

FIGURE 2.2 Absorbed photocurrent spectra for InGaAs/InAlAs MQWs as a function of applied biases.

fourth power of the QW thickness, according to the variational calculation [13], whereas the increase in size can result in a significant change in the electron-hole overlap, reducing the absorption strength of the exciton line. Therefore, there is an optimum QW thickness at which a highly efficient EA effect will arise; this is discussed in the next section. The QCSE has received a lot of attention because it is very large even at room temperature, allowing high-speed, low-driving voltage modulators and switches.

2.1.3 STRONG AND WEAK QUANTUM CONFINEMENT

From the viewpoint of practical applications to modulators, QWs operating in the long-wavelength region are important; in this region, optical fibers are transparent and their dispersion is low. The InGaAsP/InGaAsP quantum well typically used for operation in that region has less electron confinement than InGaAs/InP or InGaAs/InAlAs MQWs. This is because, in an attempt to overcome optical saturation at higher optical intensities due to hole pile-up, the QWs are designed so that the valence band offset is small. In such situations, the discrete energy level for the electron becomes quasi-bound states with a finite energy width due to the weakened confinement.

Figure 2.3 shows the calculated shapes of the absorption spectra in a QW structure under an electric field strength of 0 and 150 kV/cm, respectively, together with the wave functions [14]. In the presence of an electric field, the wave functions have an oscillatory tail that extends out of the well, as shown in the inset of Figure 2.3(a). As a result, the field-induced broadening is remarkable, i.e., the weakened confinement makes the discrete level quasi-bound with a finite width.

As mentioned above, the behavior of the excitons is very different in QWs with different barrier components even if the field is applied perpendicularly to the layers. Experimental results and calculations are shown for a 6-nm-thick InGaAsP QW with InGaAsP barriers whose photoluminescent (PL) wavelength is 1.3 μm (Figure 2.4)

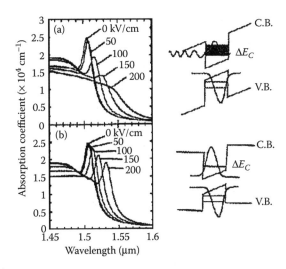

FIGURE 2.3 Variation in the calculated absorption spectra under 5 different electric field strengths with an increment of 50 kV/cm. Upper lines in (a) show the spectra of QW structure with tilt barriers, i.e., weak confinement, and lower lines in (b) show those with flat barrier, i.e., strong confinement.

and a 10-nm-thick InGaAsP QW whose PL wavelength is 1.2 μm (Figure 2.5). In the calculations, conduction-band offsets of $0.5\Delta E_g$ for the 6-nm QW structure and $0.55\Delta E_g$ for the 10-nm QW structure were assumed, where ΔE_g is the band gap energy difference. Hence, the estimated barrier heights from the ground level of the conduction band are 40 and 130 meV, respectively. As can be seen from the figures,

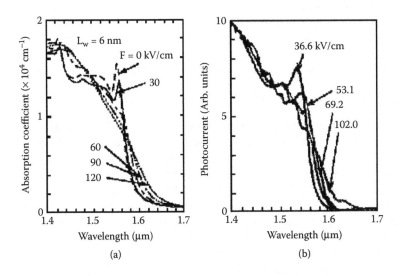

FIGURE 2.4 Comparison of the calculated absorption spectra (a) and the measured photocurrent spectra (b) for a 6-nm InGaAsP/InGaAsP single QW structure.

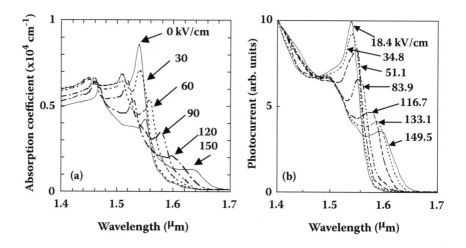

FIGURE 2.5 Comparison of the calculated absorption spectra (a) and the measured photocurrent spectra (b) for a 10-nm InGaAsP/InGaAsP single QW structure.

the barrier height difference has an important effect on the field-induced broadening, and carrier tunneling dominantly contributes to the spectral broadening. Other broadening mechanisms such as inhomogeneous field distribution and thickness and composition fluctuations play a minor role.

In contrast to InGaAsP/InGaAsP MQWs, InGaAs/InAlAs MQWs have a large ΔE_c and show strong electron confinement. Therefore, when an electric field is applied perpendicular to the layers, the exciton resonance and shifts can be observed clearly. This quantum confinement difference affects device characteristics. In particular, when tensile strain is introduced to obtain polarization-insensitive modulation, both MQWs can operate as polarization-insensitive modulators, but the power saturation level is very different between them. This is because tensile strain introduction increases valence band offset between wells and barriers, resulting in hole pile-up under high incidental optical power. The details of this will be described in the next section.

2.1.4 OPTIMIZATION OF MQW STRUCTURE

In InGaAs MQW modulators on InP substrates, the barrier and waveguide may be made from quaternary alloys of either InGaAsP or InGaAlAs. If a lattice-matched alloy with a particular bandgap is selected, the other properties are fixed at values that depend on the choice of InGaAsP or InGaAlAs. Although the bandgap discontinuity ΔE_g at the heterojunctions may be the same in both cases, the discontinuities in the conduction and valence bands ΔE_c and ΔE_v are different. For InGaAs/InGaAsP, $\Delta E_c{:}\Delta E_v$ is thought to be about 40:60 [15], whereas for InGaAs/InGaAlAs, this ratio is about 72:28 [16]. Because the effective masses of electrons and heavy holes differ by an order of magnitude, the band discontinuities have a significant effect on the well thickness appropriate for a desired output wavelength. The well thickness must be greater for InGaAlAs barriers. In the quantum-confined Stark effect (QCSE), the exciton peaks shift in proportion to the fourth power of the well thickness when the applied field is weak [17].

Moreover, ΔE_v has an effect on modulator characteristics such as frequency response and absorption saturation under high incidental optical power because of hole pile-up at the heterointerface. Especially for InGaAs/InP MQW structures with ΔE_v of 0.38 eV, the frequency response is degraded because of hole pile-up [18]. InGaAsP/InGaAsP MQW structures with a ΔE_v of 0.2, eV is used instead to reduce this degradation.

On the other hand, for the same quaternary alloys we can tailor the well thickness and well bandgap energy even though the energy band edge of the MQW is constant. Because larger absorption variations can therefore be obtained with similar applied electric fields, it is necessary to optimize the well and barrier structure for modulators.

The first proposal to improve the EA effect in MQW structures by replacing the ternary InGaAs well with a quaternary InGaAsP well was reported in 1987 [19]. The optimizations of the InGaAsP/InP [20,21] and InGaAlAs/InAlAs [20] systems have also been reported.

Quaternary QW materials are believed to be more advantageous than the conventional ternary QW materials for enhancing excitonic EA effects. Since the bandgap of quaternary materials is larger than that of ternary materials, the exciton transition energy is also greater. The QW thickness must therefore be increased for the quaternary QWs to keep the exciton transition energy the same as that for the ternary QWs. This is necessary for operating the device at a certain fixed wavelength, such as the 1.55-μm wavelength. The energy shift of quantum levels (Stark shift) induced by application of an external electric field increases with well thickness. The quaternary QW is therefore thought to have field effects more pronounced than those of the conventional ternary QWs when both are operated at the same wavelength. The increase in well thickness, however, implies a decrease in the oscillator strength of the excitonic transition. This factor may impede the above enhancement.

We define the figure of merit as $\Gamma \Delta \alpha / \Delta F$, where Γ is the optical confinement factor, and we calculate its value under the condition that it includes the field-induced broadening and the transitions between the conduction ground subband and the topmost three valence subband (1e-1hh, 1e-1lh, 1e-2hh) [21]. The former effect, which has long been neglected, is critical for practical use in weak-confinement MQWs such as InGaAsP/InGaAsP, where under a relatively weak electric field the conduction band offset is too small to confine electrons.

As a result, we obtain the optimum conditions for MQW structures, such as the QW thickness and configuration. Figure 2.6 shows the figure of merit versus QW thickness and composition. The operating wavelength is 1.55 μm and the barrier thickness is 7 nm. There are two peaks: one is for a thin well and small energy gap ($E_g = 0.7$ eV), and the other is for a thick well and large energy gap ($E_g = 0.8$ eV). As shown in Figure 2.6, the former corresponds to the first heavy-hole excitonic transition and the latter to the light-hole exciton. The former has relatively large transmission loss because of smaller detuning energy, whereas the latter has a larger figure of merit and lower transmission loss. That is, the thicker QW (about 11–12 nm) is better. This is because of the larger overlap integral of the 1e-1lh transition due to the light-hole character and to the enhanced joint density of state due to the valence band-mixing in the 1lh subband. Thicker wells also increase the optical confinement and lower the operating voltages.

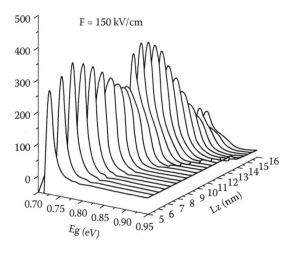

FIGURE 2.6 Figure of merit $\Gamma\Delta\alpha$ for excitonic electroabsorption effects as a function of well thickness L_z and bandgap E_g of QW materials for InGaAsP/InGaAsP QWs. The applied electric fields are 150 kV/cm.

This result gives us insight into the efficient EA effect in QWs. The introduction of tensile strain into wells, which has been used to reduce the anisotropic modulation of polarization dependence, will also affect the reduction of drive voltage according to this calculated model because the hh and lh transitions cross under a certain tensile strain, and a thicker well is used to keep the detuning energy suitable. In fact, polarization insensitive MQW modulators are operated at a lower drive voltage and high speed [22].

2.2 EA MODULATOR DESIGN PRINCIPLES

In this section, we introduce the operation principle of an EA modulator as a guide. First, the operating wavelength must be determined. The propagation loss and dispersion of silica-based single-mode fiber depend on the wavelength. We have two kinds of optical fibers whose dispersion-minimum wavelength is 1.3 μm or 1.55 μm. The former is called *normal fiber*, and the latter is called *dispersion-shifted fiber*.

Generally, the operating wavelength and the materials are not independent. For the wavelengths of low-loss region, InGaAsP or InGaAs are determined. Then we determine the waveguide thickness and its fractional content of mixture of well and barrier. Waveguide characteristics are strongly dependent on the waveguide core and cladding materials, because the optical confinement depends on the refractive index profile. The suitable detuning energy between the modulated light wavelength and the absorption band edge of EA modulators is about 40–50 meV.

For MQW structures, the thickness of an InGaAsP QW is about 7.5 nm when the well and barrier are lattice-matched to an InP substrate. If we use an InP barrier, the InGaAs well thickness is 5.0 nm when this modulator operates at 1.55 μm. Next, we determine the total thickness of the waveguide core or the MQW layer.

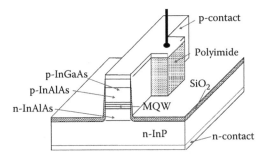

FIGURE 2.7 Schematic view of MQW EA modulator.

The barrier thickness must be thin to ensure large optical confinement and small driving voltage as far as the electron confinement is achieved. The thickness of the waveguide layer (summation of well and barrier thickness multiplied by their number) is determined from a practical point of view. That is, it depends on the operating voltage, 3-dB bandwidth, insertion loss, and so on. For typical devices, the waveguide thickness is between 0.1 and 0.2 μm, so that the operating voltage is a few volts and the 3-dB bandwidth is around 10 GHz.

A schematic diagram of an InGaAs/InAlAs MQW EA modulator is shown in Figure 2.7.

2.2.1 EA MODULATOR CHARACTERISTICS

Electroabsorption modulators have five important parameters: on/off ratio, driving voltage, 3-dB bandwidth, insertion loss, and chirping. We discuss these factors as follows.

2.2.1.1 On/Off Ratio

The most important parameter is the on/off ratio for the intensity modulator. At least 15 or 20 dB is usually needed for applying the practical systems. The on/off ratio is defined as the ratio of incident light intensity P_{in} to transmitted light intensity P_{out} and is given as

$$[\text{on/off}] = -10 \log_{10} (P_{out}/P_{in}) \text{ [dB].}$$

In EA materials, the transmitted light intensity is given by using the absorption coefficient with the incident light intensity as

$$P_{out}/P_{in} = \exp(-\Gamma \Delta\alpha L),$$

where Γ, $\Delta\alpha$, and L are the optical confinement factor, absorption coefficient change, and sample length, respectively. Therefore, the on/off ratio is given as

$$[\text{on/off}] = 0.434 \, \Gamma \Delta\alpha L \text{ [dB].}$$

2.2.1.2 Required Voltage

As shown in Figure 2.2, the absorption coefficient change $\Delta\alpha$ strongly depends on the applied voltage and the wavelength used. Based on the QCSE, the absorption coefficient peak shifts in proportion to the square of the applied electric field. In practical use, the smaller the applied voltage, the better for driving electronics circuits. Up until now, 2-V peak-to-peak operation has been necessary for high-speed (more than 10 Gb/s) operation. The applied voltage is determined by the total thickness of the i-layer, and the barrier thickness is also important. For lumped modulators, the total thickness limits the device capacitance that results in 3-dB bandwidth.

2.2.1.3 3-dB Bandwidth

The bandwidth is usually determined by the device capacitance when the device is operated in the reverse bias condition, except when it is operated by carrier injection, which is operated at very slow speed (corresponding to a few ns). When the speed is limited by the device capacitance, the 3-dB bandwidth is given as

$$[f_{3dB}]_1 = 1/\pi RC,$$

where R and C are the load resistance and capacitance, respectively. The device capacitance is proportional to sample length L and width W, and it is inversely proportional to i-region thickness d when the stray capacitance is neglected and the electric field is applied perpendicularly to the i-region. The last assumption is discussed in the last section, where a new structure of lateral p-i-n using a parallel field is proposed. As the capacitance decreases, the frequency response increases. Usually, stray capacitance induced from the bonding pad cannot be neglected and some countermeasures, such as polyimide with small dielectric constant, are used to reduce the stray capacitance.

The above discussion is based on the assumption that the sample is short enough and the transit time of light through the sample is very short. When we use long samples, the bandwidth is limited by the transit time τ of the light through the sample and the bandwidth is given as

$$[f_{3dB}]_2 = 1.39/\pi\tau = 1.39c_0/\pi n_g L,$$

where n_g is the refractive index of the waveguide and c_0 is the speed of light in a vacuum. As an example, we calculate this value for InGaAsP with a refractive index of 3.24 as

$$[f_{3dB}]_2 L = 4.13 \text{ GHz cm.}$$

That is, the 3-dB bandwidth for the lumped InGaAsP device is limited to above $[f_{3dB}]_2$ even if $[f_{3dB}]_1$ is increased.

To improve this limitation, a traveling-wave-type structure is proposed and demonstrated. In this structure, a traveling line is arranged along the optical waveguide to match the transmission velocity of microwave and optical waves. In this case, the bandwidth is [23]

$$[f_{3dB}] = 1.39c_0/\pi[n_g - n_m]L,$$

where n_m is the refractive index of the transmission line for the modulated microwave. Based on this equation, $[f_{3dB}]$ seems to increase monotonously if we increase L when the velocity matching condition is fulfilled. Practically, the losses for optical and microwave transmission are not neglected; however, the maximum value in the reported data is 350 GHz, which was obtained by using superconducting materials for the ohmic contact [24].

2.2.1.4 Insertion Loss

Insertion loss consists of transmission loss, reflection loss, and coupling loss. The transmission loss consists of residual absorption loss of the material, free carrier loss, and scattering loss. The coupling loss is due to the mode-spot size mismatch between the incident light and the guided light. The reflection loss is at most 3 dB and is eliminated by antireflection coating.

The transmission loss α_{tran} consists of the inherent absorption α_g due to the energy difference between the incident light and the absorption edge of the waveguide material, free carrier absorption α_{fc} due to the free carrier in the waveguide and the cladding layer, and light scattering α_s, and is given by

$$\alpha_{tran} = \Gamma\alpha_g + (1 - \Gamma)\alpha_{fc} + \alpha_s,$$

where Γ is the optical confinement factor (Figure 2.8). This value depends on the refractive index profile. The free carrier absorption in the p-doped cladding layer is much larger than that in the n-doped cladding layer, by as much one order of magnitude for InP. The doping level is required to be small to provide low transmission loss, while high doping is required so that the applied voltage is biased only to the i-region, resulting in small bias voltage. Therefore, the p-doping is chosen to be 5×10^{17} to 1×10^{18} cm^{-3}.

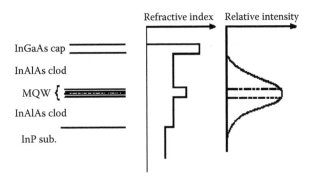

FIGURE 2.8 Electric field distribution and optical confinement factor Γ for an InGaAs/InAlAs MQW modulator three-layer waveguide.

In a semiconductor waveguide, the mode spot size is generally very small, especially in the direction perpendicular to the layer, due to the large refractive index difference between the guide layer and the cladding layer. This strong optical confinement produces small driving voltage, causing effective overlap between the optical field and the electric field which results in small spot-size. Therefore, the large insertion loss in semiconductor optical devices is mainly due to the coupling loss; and some counter-plans, such as monolithic integration with laser diodes or mode-spot size transformers, have been tried in an attempt to minimize this loss.

2.2.1.5   Chirping

Any change of absorption coefficient in a material structure will be accompanied by a phase shift of light, since the real and imaginary parts of a dielectric constant constitute a Kramers-Kronig transform pair. This causes frequency chirping in intensity modulation and intensity fluctuation in phase modulation. When chirping is induced, the transmitted optical pulses are broadened through the fiber due to the dispersion. Though the chirping for external modulators is thought to be much smaller than that for direct-modulated laser diodes, it is a limiting factor for the capacity of long-haul high-bit-rate optical-fiber communication systems due to fiber dispersion, as shown in Figure 2.9.

The magnitude of chirping is defined as the ratio of the change of refractive index Δn to the change of the extinction coefficient Δk. It has been proposed that waveguide Mach-Zehnder amplitude modulators may be operated in a perfectly chirpless mode. Devices designed to operate chirp-free have been reported, whereas some chirping exists for EA modulators; however, its magnitude is at most 1.5 based on previously reported data (Table 2.1).

However, the lowest fiber-dispersion penalty is generally not obtained for a frequency chirp parameter equal to zero. It can be advantageous to choose a nonzero value for the chirp parameter, depending on the fiber dispersion coefficient and distance, to provide some amount of pulse compression.

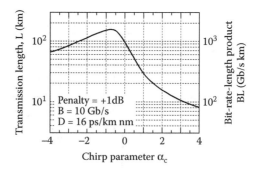

FIGURE 2.9 Calculated transmission length versus chirping parameter. Modulation speed, power penalty, and fiber dispersion are assumed to be 10 Gbit/s, 1 dB, and 16 ps/km nm, respectively.

TABLE 2.1
Review of α Parameter for EA Modulators

Author	Firm	Material	α Parameter	Detuning Energy, meV	Ref.
Y. Noda	KDD	InGaAs/InP F-K	1–2	51.5	JLT, LT-4, 1445, '87
T.H. Wood	AT&T	GaAs/AlGaAs QCSE	1	12 (TM)	APL, 50, 798, '87
K. Wakita	NTT	InGaAs/InAlAs QCSE	0.6–0.7	22–27	JJAP, 26, 1169, '87
H. Soda	Fujitsu	InGaAsP/InP F-K	1	68	EL, 24, 1194, '88
T. Saito	NEC	InGaAsP/InP F-K	0.6	85.7	OEC '90, 13A2-4, '90
K. Wakita	NTT	InGaAs/InAlAs QCSE	0.7	27	PTL, 3, 138, '91
M.S.Whalen	AT&T	InGaAs/ InP QCSE	0.6–0.8	24.7–28.3	PTL, 3, 451, '91
M. Suzuki	KDD	InGaAsP/InP F-K	0.2–0.4[a]	56–71	PTL, 3, 451, '91
J. Langanay	Alcatel	InGaAsP/InGaAsP QCSE	−0.2–0.2	24.7–27.4	APL, 62, 2067, '93
F. Devaux	CNET	InGaAsP/InGaAsP QCSE	1.2	—	PTL, 4, 720, '92
F. Koyama	TIT	InGaAs/InGaAsP QCSE	0.4–1	5.4–21.2	PTL, 5, 1389, '93
I. Kotaka	NTT	InGaAsP/InGaAsP QCSE	0.8	30.5	PTL, 5, 61, '93
F. Devaux	CNET	InGaAsP/InGaAsP QCSE	−2.0–3.0	—	PTL, 3, 1288, '93
T. Kataoka	NTT	InGaAsP/InGaAsP QCSE	0.2–1.4	30.5	EL, 30, 872, '94
T.H. Wood	AT&T	InGaAsP /InP QCSE	0.5	28.3	JLT, LT-12, 1152, '94
T. Ido	Hitachi	InGaAs/InAlAs QCSE	0.6	32.2	PTL, 6, 1207, '94
J. Shimizu	Hitachi	InGaAlAs/InGaAlAs QCSE	−2.5–1.0	26.7	EL, 38, 821, '02
Y. Miyazaki	Mitsubishi	InGaAsP/InGaAsP QCSE	−0.7–0.7	—	J.QE., 38, 1075, '02

[a] Underestimated; revised values 0.585–0.795(JLT, LT-12, '94).

The chirp parameter for the EA modulator is shown in Figure 2.10, where the transverse axis indicates the transmission loss with pre-bias. Note that the chirp parameter decreases as the operating wavelength approaches the absorption edge energy, while the absorption at zero bias (transmission loss) increases. That is, the chirp parameter and the transmission loss contradict each other, and some compromise is necessary. In general, the transmission loss is too large to be used under the condition of negative chirp parameter. Recent advanced technologies produce highly efficient optical amplifiers and monolithically integrated light sources with EA modulators and DFB lasers, which enable us to use EA modulators with large propagation loss. The recent improvements of the chirping in the external modulators will be discussed in the last section.

2.2.1.6 Figure of Merit

As described in the previous sections, there are trade-offs among the above five parameters. For example, on/off ratio is proportional to the product of absorption coefficient change and sample length L, whereas 3-dB bandwidth f_{3dB} is inversely proportional to L and proportional to i-region thickness d, and the operating voltage is proportional to d.

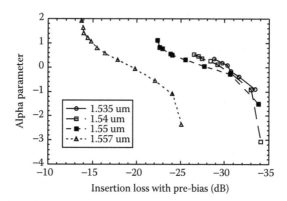

FIGURE 2.10 Chirp parameter versus insertion loss as a function of incident light wavelengths.

Moreover, high-speed and low-driving-voltage operation is a typical trade-off relationship, because device speed is usually limited by capacitance (for lumped modulators) that is associated with the guide layer thickness, and the thickness limits the operating voltage. The on/off ratio is given by the product of the absorption coefficient change and the sample length. Therefore, maximization of the absorption coefficient change associated with applied voltage is necessary. For waveguide structures, the optical confinement factor must be taken into account. For MQW structures, thick QW and thin barriers are effective for this purpose as far as the quantum size effect can work. Tensile-strained quantum wells can reduce the operating voltage drastically. This introduction can also achieve both polarization insensitivity and low chirp [22].

Figure 2.11 shows the summary of the required RF power vs. capacitance-limited 3-dB bandwidth for recently developed modulators. A drive impedance of 50 Ω is

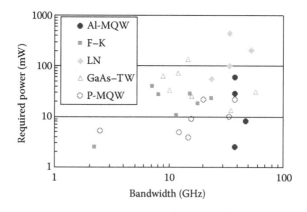

FIGURE 2.11 Required power versus capacitance-limited bandwidth for recently developed modulators. Solid squares, F-K, Franz-Keldysh; open triangles, LiNbO$_3$ modulator; circle with bar, solid diamond circle within circle, open circle modulator; solid circle, InGaAs/InAIAs MQW. Required drive power $P_{ac} = V_{ac}^2/50$, $V_{ac} = V_{pp}/2\sqrt{2}$, $V_{pp} = V_\pi$ or V_{10dB}.

assumed, and the operating voltage is the total voltage swing for either a π-phase shift for a phase modulator or 10 dB on/off ratio for an EA modulator. F-K stands for Franz-Keldysh modulators, TW indicates traveling waveguide structure modulators [25–28], respectively, and the triangles indicate LiNbO$_3$ modulators [29,31]. Reduction of device capacitance results in high-speed operation, whereas drive voltage increases. The ratio of device voltage to bandwidth figure of merit for InGaAs/InAIAs MQW modulators is lower than for any other existing optical modulators.

Recently, a well-established approach to obtain very wide-band modulators, the so-called traveling-wave design, has been reported [26,33,34]. In this design, the electrode is designed as a transmission line so electrode capacitance is distributed and does not limit the modulator speed due to constant limitation of RC time. Although this method has been applied for a long time to LiNbO$_3$ and semiconductor phase modulators because of their relatively long waveguide length, it has been applied to shorter EA modulators as well. Details will be discussed in the last section of this chapter.

2.3 MONOLITHIC INTEGRATION OF EA MODULATORS AND DFB LASERS

We saw in Section 2.2 that semiconductor modulators have a large insertion loss, mainly in the form of coupling loss because of their small spot size. One approach to reducing the coupling loss is to integrate the modulators with light sources. In this section, we consider the monolithic integration of modulators with light sources.

Monolithic integration of opto-electronic devices has been studied for many years. Since the first concept of OEICs (opto-electronic integrated circuits) was proposed by Miller in 1969 [35], a great deal of research on developing OEICs has been done in many laboratories. Though the OEICs concept incorporates advanced technical research and development, and many devices have been fabricated, practical applications have been realized for only a few devices. One exception, however, is monolithic integration of modulators and laser diodes. This integration has many advantages: it offers low insertion loss, high output power, low chirping, and, by reducing the number of parts, increased reliability. This should reduce the overall system cost. Usually the operating light wavelength for a laser is different from that for a modulator, and the former is forward-biased whereas the latter is reverse-biased. Therefore, the two devices need to be isolated both electrically and optically.

Many attempts to integrate modulators and laser diodes have been reported. However, it was not until the development of DFB lasers that success in integrating modulators and lasers was demonstrated [36, 37], because conventional Fabry-Perot lasers needed optical cavities and were never isolated optically from modulators. Since then, integrated light sources using the EA effect have been reported [38–43] with the emphasis on their ease of fabrication. In particular, a novel yet simple method for monolithic integration of a DFB laser and an EA modulator operating in the 1.5-μm window has been proposed and demonstrated [38]; in this method, the two devices are made of a single active layer.

2.3.1 MONOLITHIC INTEGRATION OF FRANZ-KELDYSH MODULATORS AND DFB LASERS

Since the first demonstration of an integrated light source was reported in 1987, many laboratories in Japan have worked to develop this technology. System experiments using integrated light sources [42–46] have been reported, and laboratories in other countries have begun to investigate their use in multigigabit long-haul transmission systems [47,48]. Figure 2.12 shows another integrated light source [40]. The power emitted from the modulator facet is 17 mW and the 3-dB bandwidth is over 10.3 GHz. Limited wavelength broadening (chirping) within 0.01-nm under 10-Gbit/s NRZ modulation was also demonstrated. As an accelerated life test, under 70°C ambient temperature and high-output power operation, the integrated light source showed stable operation over 3000 hours without any degradation. The device voltage of EA modulators utilizing the Franz-Keldysh effect was still too high for high-speed IC drivers and it was necessary to incur increased insertion loss to reduce drive voltage because of the small detuning energy. At present, this integrated light source is suitable for systems with relatively low modulation speed of about 2.4 Gbit/s.

2.3.2 MONOLITHIC INTEGRATION WITH MQW MODULATORS

Optical modulators made with MQW structures can produce excellent performance because of their large EA effect (QCSE). DFB lasers with lower threshold current densities, improved high frequency response, and narrower linewidth have been developed by introducing MQW structures in place of the bulk in the active layer [49]. Here we describe a low drive-voltage and high-speed monolithic light source with an MQW structure for both the DFB laser and the external modulator. At first, the MOVPE/MBE hybrid growth technique was used for the integration [50]; today, the MOVPE technique is more popular.

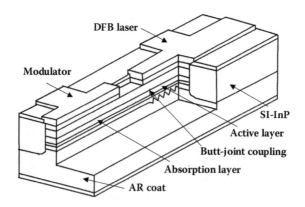

FIGURE 2.12 Schematic drawing of a DFB laser/electroabsorption modulator integrated light source.

2.3.3 ADVANCES FOR MONOLITHIC INTEGRATION

The previous hybrid method of monolithically integrating an MQW modulator and a DFB laser is too complicated to permit easy fabrication. However, easy and simple methods using a vertical mode coupling structure [51,52] and/or a simultaneous selective area growth (SAG) method [42,43,53–57] have been reported.

Figure 2.13 shows the schematic diagram of the vertical mode coupling integrated light source [52]. In this case, the two MQW core layers (the upper for a laser and the lower for a modulator) are stacked on top of each other in one growth step, separated by a thin InP spacer layer. After the grating on the waveguide layer is fabricated, the upper core to the etching stop layer is etched off selectively by wet chemical etching and the cladding layer is regrown. The crystal quality and optical coupling efficiency are improved by the regrowth on the flat layer and two-core structure, compared with those of the butt-joint-coupling method. The main advantage remains the independent optimization of the two elements.

Recently, simultaneous selective area growth has been reported [53–57]. With this method, both the laser section and the modulator section can be grown simultaneously by using MOVPE, based on the difference in growth speed between the dielectric mask widths. Optical coupling efficiency is almost 100% and the fabrication process is easier than butt-joint processing, which has been studied for a long time. One potential drawback, however, may lie in the difficulty of independent optimization of each component. The reported device characteristics have been fair and transmission experiments were demonstrated at 2.5 Gbit/s over 517 km [54] and 600 km [55] of standard fiber.

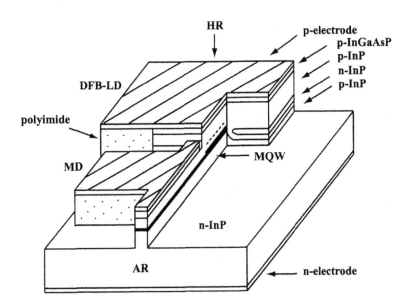

FIGURE 2.13 Schematic drawing of a strained InGaAsP/InGaAsP MQW DFB laser/ electroabsorption modulator integrated light source and its cross-section; (a) and (b) indicates laser side and modulator side, respectively.

Moreover, butt-joint coupling between an MQW laser active layer and an MQW modulator waveguide has been reported and 10-Gb/s, 120-km transmission has been achieved in one experiment [56]. The butt-joint coupling has previously been considered to be too difficult for applying MQW structures due to the excessive growth at the interface. That is, thick quantum wells will produce a small energy band gap causing increasing transmission loss. However, by using MOVPE techniques, high-coupling efficiency and high-emitting power from the facet of the modulator have been obtained and over 40-Gb/s modulation has been demonstrated [59–61].

2.4   RECENT NEW APPLICATIONS OF EA MODULATORS

In this section, we discuss recent research and development for EA modulators: ultrahigh-speed modulation, high allowability of incidental optical power, and low chirping modulation.

2.4.1   ULTRAHIGH-SPEED OPERATION

Low-driving-voltage operation is the key point bringing high-speed (more than 40 GHz) modulators into practical use because it eliminates the need for high-speed electrical amplifiers. There is generally a trade-off between modulator speed and driving voltage for lumped modulators. One way to achieve high-speed operation without increasing voltage is to use short samples with small capacitance as shown in Figure 2.14 [58,62,63], where an MQW modulator is integrated with nondoped waveguides. This structure enables us to fabricate samples less than 100 μm long, but the extinction ratio decreases. Table 2.2 shows the high-speed modulators with modulation bandwidth of over 40 GHz reported so far [34,58–66]. As mentioned in the previous section, there is an optimum condition at which EA modulators operate both at low driving and high speed, and our conclusion is that a well thickness of 12 nm is the best when we introduce tensile strain in the wells [22].

The introduction of tensile strain enhances the EA effect, resulting in reduced drive voltage. This is due to heavy- and light-hole degeneracy, where the absorption of heavy-hole excitons transition coincides with that of the light-hole, as well as to

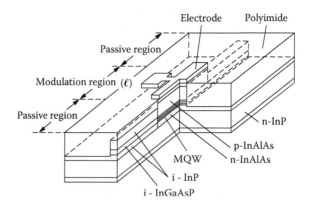

FIGURE 2.14 Structure of an MQW modulator with integrated waveguides.

TABLE 2.2
Reported High-Speed Semiconductor Modulators

Year/ Month	3dB (GHz)	Ext. Ratio, Drive Voltage	L(μm)	Material	Affiliation	Ref.
1990.5	40	20 dB, 7 V	100	InGaAs/InAlAs MQW	NTT	CLEO
1993.10	38	10 dB, 8 V	1	GaAs/AlGaAs MQW	UCSB	DRC
1994.3	35	13 dB, 6 V	3 mm	InGaAs/InP bulk	HHI	OFC
1994.8	35	19 dB, 4.2 V	120	InGaAsP/InGaAsP MQW	CNET	EL, 30, 1347
	28	10 dB, 2.5 V	115			
1995.2	40	10 dB, 2.6 V	50	InGaAs/InAlAs MQW	Hitachi	PTL, 7, 170
	30	20 dB, 2.6 V	100			
1995.3	40	$V_\pi = 25$ V	2 cm	GaAs/AlGaAs bulk	UCSB	OFC
	(70)	10 V				
1995.7	50	20 dB, 3.4 V	63	InGaAs/InAlAs MQW	Hitachi	IOOC
	42	19 dB, 4.4 V	75	(tensile)	CNET	IOOC
1995.9	42	20 dB, 2 V	107	InGaAs/InAlAs MQW	NTT	ECOC
1997.8	50	15 dB, 9 V	200	InGaAs MQW	NTT	EL, 33, 1580
2000.7	50	20 dB, 3 V	100	InGaAsP MQW	Oki	OECC
2001.3	70	20 dB, 3 V	225	InGaAsP MQW	NTT	EL, 37, 1
2001.3	40	10 dB, 3 V	75	InGaAsP MQW	Mitsubishi	OFC

the electric field effect difference in the absorption edge shift and oscillator strength decrease between them [67]. This enhances the oscillator strength at the zero-bias condition and absorption coefficient change. Based on the data, a strain of −0.45% gives the best figure of merit. Polarization insensitivity was achieved and extinction ratio was about 20 dB at the swing voltage of 1.0 V. The frequency response of this modulator was measured. The ratio of 3-dB bandwidth to drive voltage required for a 20-dB on/off ratio is over 40 GHz/V. Clear eye opening of 40 Gbit/s was obtained for a 0.9 Vpp pseudo-random modulation signal, and 1.5V DC bias was observed. This modulator was used in a 320-km long dispersion-shifted fiber transmission experiment with four-channel wavelength division multiplexing, in the first experiment of its kind [68]. In this experiment, no power penalty was observed. This indicates that the strained MQW modulator gets an advantage over power-hungry LiNbO$_3$ modulators over 40 Gbit/s without a high-speed electrical amplifier.

Another way to achieve high-speed operation without increasing voltage is to use traveling-wave-electrode configuration, where the device speed is free from capacitances and over 40 Gbit/s large-signal modulation has been reported [30,60]. Figure 2.15 shows a schematic design of a TW-EA modulator integrated with a DFB laser. However, their inevitably small characteristic impedance (about 25 Ω) is much smaller than that of the standard 50-Ω rf connections. A novel traveling-wave modulator electrically matched for InP-HBT drivers was demonstrated, and a 40-Gbit/s, 2-km SMF transmission with a 0.3-dB penalty at a 1.3-μm wavelength was reported [69].

A new electroabsorption modulator consisting of a lateral p-i-n configuration whose i-region is composed of multiple quantum wells (MQWs) has been proposed

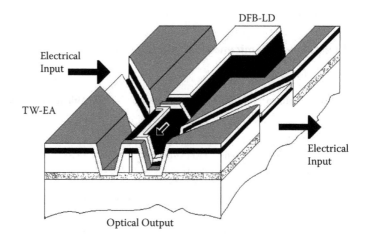

FIGURE 2.15 Schematic diagram of TW-EA-DFB.

[70] as shown in Figure 2.16; the device characteristics, polarization insensitivity, and low insertion loss are simulated for high-speed, low-driving-voltage operation. The 3-dB bandwidth and the driving voltage required to achieve a 20-dB extinction ratio are estimated to be over 250 GHz and less than 2V, respectively, assuming that the speed is limited by the device capacitance. This structure enables us to apply a parallel field for MQW layers. The exciton resonance can easily broaden and disappear with a smaller field intensity than that of a perpendicular field (QCSE) [12]. On the other hand, the i-region width between an n- and p-doped region can be designed to optimize the field intensity and coupling efficiency between an optical fiber and the modulator. This is another characteristic for this structure, because the previous structure (see Figure 2.7) has limited thickness of MQWs for low driving voltages that results in large device capacitance and limited optical confinement and coupling efficiency between an optical fiber and the modulator.

2.4.2 HIGH INCIDENTAL OPTICAL POWER OPERATION

Another topic of current research is high allowability of incidental optical power. At present, the incidental optical power increases with the need for dense wavelength

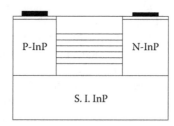

FIGURE 2.16 Schematic diagram of a lateral p-i-n MQW modulator.

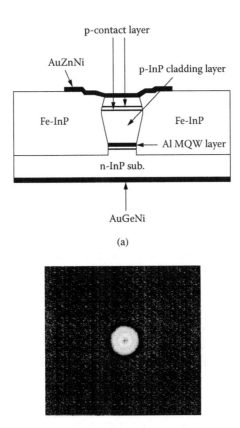

(a)

(b)

FIGURE 2.17 A cross-section of a fabricated modulator buried in semi-insulating InP (a), and its near field pattern (b).

division multiplexing (DWDM) systems; however, there are few reports on this topic. It was confirmed that the upper limit of allowability is determined by the product of absorbed photocurrent and applied voltage. The value of an EA modulator with a semi-insulating buried heterostructure (SIBH) (see Figure 2.17) is superior to that of the high-mesa structure that almost all reported discrete EA modulators are made from [71]. Figure 2.18 shows the relationship between breakdown voltage and absorbed photocurrent. Compared with high-mesa structures, the SIBH can withstand high-input power and is not observed to deteriorate even at the greatest experimentally obtained power level (20 dBm). This is considered to be due to the difference of thermal conductivity/capacity between the two structures, and SIBH modulators are expected to have high reliability.

It has been a serious problem for a long time that modulation characteristics of the usual Fe-doped SIBH EA modulators are inferior to those of high-mesa modulators because of the interdiffusion between the transition metal (Fe) and the zinc (p-type dopant in the cladding layer). Ruthenium-(Ru) doped SIBH modulators have been developed [72] and their modulation characteristics such as extinction ratio,

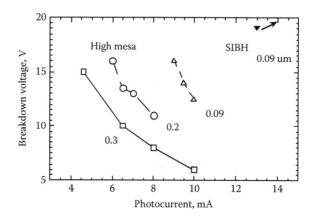

FIGURE 2.18 Breakdown voltage versus absorbed photocurrent for high-mesa structure and SIBH modulators with MQW guide thickness as a parameter. Incidental light is TE-polarized with a wavelength of 1.55 μm.

frequency response, and dark current are superior to those of Fe-doped SIBH modulators and similar to those of high-mesa structures.

In EA modulators, photocarriers are generated inevitably, and piling up of the carriers is induced by high-input power degrades rf response. To reduce the density of photocarriers, a small valence-band offset and thin barrier have been designed as discussed in Section 2.1.3 and Section 2.1.4. A strain-compensated structure with a compressive InGaAsP QW and a tensile InGaAsP barrier to provide small valence band discontinuities has been proposed to reduce the hole escape time from QWs [73] at the expense of polarization insensitivity. This choice is permitted when applying this technology to the application of monolithically integrated light sources. Moreover, the introduction of InGaAlAs barrier [74] or tensile-strained asymmetric QW [75] has been proposed and demonstrated for this purpose. These trials have been made possible thanks to a better understanding of the chirping behavior of these structures, as discussed in the following section.

2.4.3 CHIRPING REDUCTION FOR EA MODULATORS

The primary factor for developing external modulators is considered to be a strong demand for devices that operate with low or negative chirp. At present, low-chirp operation has been achieved at the expense of increased insertion loss and reduced extinction ratio [76]. Although polarization-insensitive, strain-compensated InGaAlAs/InGaAlAs MQW modulators have been developed, negative-chirp operation without increased transmission loss has not yet been accomplished. Based on the applied bias dependence of chirp parameter α for lattice-matched QWs with 20-nm thickness, the α is almost zero even at zero bias [22], although the quantum size effect is small. This indicates that poor electron and/or hole confinement will give negative α parameters. In fact, negative α parameters have been reported for bulk modulators based on

Franz-Keldysh effect (no QCSE) even zero bias [77]; however, the detuning energy was small (about 28 meV) and insertion loss was large (about 15 dB). Various trials for reducing the escape time of photocarriers and the degradation of frequency response have been done as discussed in the previous section; these result in reducing the α parameters. However, at present the α parameter is still larger than desired under conditions of nonzero bias. More investigation is necessary.

2.4.4 OPTICAL GATE INTEGRATING A UTC-PD AND TW-EA MODULATOR

Ultrashort optical pulses synchronized to an electrical clock are required in ultrahigh-bit-rate soliton transmission and optical signal processing systems. When we apply a large signal to the modulator, the transmitted light is narrower than half the width of the driving sinusoidal wave, and short optical pulses can thus be obtained under a high-frequency operation [78]. The width depends on the repetition frequency and driving voltage to the extent that the modulator can respond to the driving frequency. Low-driving-voltage EA modulators enable us to directly drive optical signals by a high-output uni-traveling carrier photodiode (UTC-PD) [79] without electrical amplifiers. This optical gate demonstrated a 40-Gbit/s 4:1 demultiplexing operation [80]. Recently, monolithic integration of a UTC-PD and an EA modulator has been reported and demonstrates a 160 Gbit/s 16:1 demultiplexing operation [81] and a 320-Gbit/s 32:1 demultiplexing operation [82]. Figure 2.19 shows the layout of PD–EA modulator, and good eye opening was observed for the eye diagrams of the input 160- and 320-Gbit/s data, respectively, and demultiplexed 10-Gbit/s data. An optical gate integrating a UTC-PD and TW-EA modulator will be one of the key devices for signal processing in high-bit-rate TDM channels.

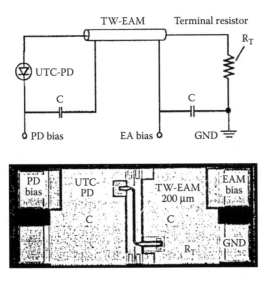

FIGURE 2.19 Equivalent circuit and layout of PD-EA modulator.

REFERENCES

1. K. Wakita, *Semiconductor Optical Modulators*, Kluwer Academic Publishers, 1998. Contains detailed references of semiconductor modulators up to 1995.
2. W. Franz, *Z. Naturforsch.*, 13a, 484, 1958.
3. L. V. Keldysh, *Zh. Eksp. Teor. Fiz.*, 34, 1138, 1958 [*Sov. Phys.—JETP*, 7, 788, 1958.
4. N. Bottka and L. D. Hucheson, *J. Appl. Phys.*, 46, 2645, 1975.
5. J. C. Campbell, J. C. DeWinter, M. A. Pollack, and R. E. Nahory, *Appl. Phys. Lett.*, 32, 471, 1978.
6. N. K. Dutta, and N. A. Olessen, *Electron. Lett.*, 20, 634, 1984.
7. Y. Noda, M. Suzuki, Y. Kushiro, and S. Akiba, *J. Lightwave Technol.*, LT-4, 1445, 1986.
8. H. Soda, K. Nakai, H. Ishikawa, and H. Imai, *Electron. Lett.*, 21, 1232, 1987.
9. T. Okiyama, I. Yokota, H. Nishimoto, K. Hironishi, T. Horimatsu, T. Touge, and H. Soda, *ECOC' 89*, MoA1-3, 1989.
10. N. Henmi, S. Fujita, T. Saito, M. Yamaguchi, M. Shikada, and J. Namiki, in *Digest of Conference on Optical Fiber Communication, OFC '90*, 1990, Paper No. PD8-1.
11. G. Mak, C. Rolland, K. E. Fox, and C. Baauw, *Photon. Technol. Lett.*, 2, 10, 1990.
12. D. A. B. Miller, D. S. Chemla, T. C. Damen, A. C. Gossard, W. Wiegmann, T. H. Wood, and C. A. Burrus, *Phys. Rev.*, B32, 1043, 1985.
13. G. Bastard, *Phys. Rev.*, B30, 3547, 1985.
14. T. Yamanaka, K. Wakita, and K. Yokoyama, *Appl. Phys. Lett.*, 65, 1540, 1995.
15. B. Broberg, and S. Lindgren, *J. Appl. Phys.*, 55, 3376, 1984.
16. S. R. Forrest, P. H. Schmidt, P. B. Wilson, and M. L. Kaplan, *Appl. Phys. Lett.*, 45, 1199, 1984.
17. G. Bastard, E. E. Mendez, L. L. Chang, and L. Esaki, *Phys. Rev.*, B28, 3241, 1983.
18. U. Koren, B. I. Miller, T. L. Koch, G. Eisenstein, R. S. Tucker, I. Bar-Joseph, and D. S. Chemla, *Appl. Phys. Lett.*, 51, 1132, 1987.
19. H. Temkin, D. Gershoni, and M. Panish, *Appl. Phys. Lett.*, 50, 1985.
20. S. Nojima and K. Wakita, *Appl. Phys. Lett.*, 53, 1958, 1988.
21. T. Yamanaka, K. Wakita, and K. Yokoyama, *InP and Related Materials*, IPRM95, ThP46, May 9–13, 1995, Sapporo, Hokkaido, Japan, p. 544.
22. K. Wakita, *SPIE 97*, 3038, 39, 1997.
23. A. Yariv, *Quantum Electronics*, 2nd ed., John Wiley & Sons, 1975, p. 347.
24. J. Nees, S. Williamson, and G. Mourou, *Appl. Phys. Lett.*, 54, 196–1964, 1989.
25. R. G. Walker, *IEEE J. Quantum Electron.*, 27, 654, 1991. Information can also be found in the GEC-Marconi catalog.
26. N. Dagli, *IEEE Trans.* Microwave Theory Tech., 47, 1151, 1999. Contains references of wide-band modulators up to the time of publication.
27. R. Spickermann, N. Dagli, and M. G. Peters, *Electron. Lett.*, 31, 915, 1995.
28. R. Spickermann, S. R. Sakamoto, M. G. Peters, and N. Dagli, *Electron. Lett.*, 32, 1095, 1996.
29. O. Mitomi, K. Noguchi, and H. Miyazawa, *IEEE Trans. Microwave Theory Tech.*, 43, 2203, 1995.
30. K. Noguchi, O. Mitomi, and H. Miyazawa, IEEE *J. Lightwave Technol.*, 16, 615, 1998.
31. R. C. Alferness, *IEEE Trans. Microwave Theory Tech.*, MTT-30, 1121, 1982.
32. H. H. Liao, X. B. Mei, K. K. Loi, C. W. Tu, P. M. Asbeck, and W. S. C. Chang, *SPIE. Optoelectron. Integrated Circuits*, Vol. 3006, 291, 1997.
33. K. Kawano, M. Kohtoku, M. Ueki, T. Ito, S. Kondoh, Y. Noguchi, and Y. Hasumi, *Electron. Lett.*, 33, 1580, 1997.

34. S. Z. Zhang, Y. J. Chiu, P. Abram, and J. E. Bowers, *IEEE Photon. Technol. Lett.*, 11, 191, 1999.

35. S. E. Miller, *Bell Syst. Tech. J.*, 48, 2059, 1969.

36. Y. Kawamura, K. Wakita, Y. Itaya, Y. Yoshikuni, and H. Asahi, *Electron. Lett.*, 22, 242, 1986.

37. Y. Kawamura, K. Wakita, Y. Yoshikuni, Y. Itaya, and H. Asahi, *IEEE J. Quantum Electron. Lett.*, QE-23, 915, 1987.

38. A. Ramdane, F. Devaux, N. Souli, D. Delprat, and A. Ougazzaden, *IEEE J. Quantum Electron.*, 2, 326, 1996. Contains references for monolithic integration of MQW lasers and modulators up to the time of publication.

39. M. Suzuki, Y. Noda, H. Tanaka, S. Akiba, Y. Kushiro, and H. Isshiki, *IEEE Lightwave Technol.*, LT-5, 1277, 1987.

40. H. Soda, K. Nakai, and H. Ishikawa: *ECOC'88, Technical Digest*, 1988, Brighton, p. 227.

41. N. Henmi, S. Fujita, T. Saito, M. Yamaguchi, M. Shikata, and J. Namiki, *OFC '90*, PD8-1, 1990.

42. T. Kato, T. Sasaki, N. Kida, K. Komatsu, and I. Mito, in *17th European Conference on Optical Communication, ECOC'91*, Paper No. WeB7-1.

43. M. Aoki, M. Takahashi, M. Suzuki, H. Sano, K. Uomi, T. Kawano, and T. Takai, *Photon. Technol. Lett.*, 4, 580, 1992.

44. T. Saito, N. Hemmi, S. Fujita, *OEC'90, Tech. Dig.*, 13A2–4, 1990.

45. T. Okiyama, I. Yokota, H. Nishimoto, K. Hironishi, T. Horimatsu, T. Touge, and H. Soda, *ECOC' 89*, MoA1–3, 1989.

46. M. Suzuki, H. Tanaka, H. Taga, S. Yamamoto, and Y. Matsushima, *J. Lightwave Technol.*, 10, 90, 1992.

47. P. Ojala, C. Pettersson, B. Stoltz, A. -C. Morner, and M. Janson, *ECOC' 93*, WeC7.4.

48. M. Gibbon, G. H. B. Thompson, D. Boyle, J. P. Stagg, B. Patel, D.J. Moule, S. Wheeler, E. J. Thrush, K. Warbrich, and P. Gurton, *ECOC '93*, WeC7.5.

49. Y. Arakawa and A. Yariv, *IEEE J. Quantum Electron.*, QE-21, 1666, 1985.

50. K. Wakita, I. Kotaka, H. Asai, M. Okamoto, and Y. Kondo, *Photon. Technol. Lett.*, 4, 16, 1992.

51. U. Koren, B. Glance, B. I. Miller, M. G. Young, M. Chien, T. H. Wood, L. M. Ostar, T. L. Koch, R. M. Jopson, J. D. Evankow, G. Raybon, C. A. Burrus, P. D. Magill, and K. C.Reichmann, in *Tech. Dig. Opt. Fiber Communication Conf. (OFC '92)*, San Jose, CA, 1992, p.124.

52. K. Sato, I. Kotaka, K. Wakita, Y. Kondo, and M. Yamamoto, *Electron. Lett.*, 29, 1087, 1993.

53. E. Zielinski, D. Bauns, H. Haisch, M. Klenk, E. Satzke, and M. Schilling, in *Proc. InP and Related Compounds*, 1994, ME-1, p. 68.

54. J. E. Johnson, T. Tanbun-Ek, Y. K. Chen, D A. Fishman, P. A. Morton, S. N. G. Chu, A. Tate, A. M. Sergent, P. F. Sciortino, and K. W. Wecht, *IEEE Semiconductor Conf.*, 1994, M. 47.

55. K. C. Reichmann, P. D. Magill, U. Koren, B. I. Miller, M. Young, M. Newkirk, and M. D. Chien, *Photon. Technol. Lett.*, 5, 1093, 1993.

56. K. Morito, R. Sahara, K. Sato, Y. Kotaki, and H. Soda, *Electron. Lett.*, 31, 975, 1995.

57. J. Shimizu, M. Shirai, T. Tsuchiya, A. Taike, and S. Tsuji, *Tech. Dig. CLEO/Pacific Rim 2001*, Chiba, Japan, 2001, Vol. 2, p. 54.

58. T. Ido, S. Tanaka, M. Suzuki, M. Koizumi, H. Sano, and H. Inoue, *J. Lightwave Technol.*, 14, 2026, 1996.

59. H. Takeuchi, K. Tsuzuki, K. Sato, M. Yamamoto, Y. Itaya, A. Sano, M. Yoneyama, and T. Otsuji, *Photon. Technol. Lett.*, 9, 572, 1997.

60. Y. Akage, K. Kawano, S. Oku, R. Iga, H. Okamoto, Y. Miyamoto, and H. Takeuchi, Proc. ECOC 2000, Munich, PD004, 2000; Y. Akage, K. Kawano, S. Oku, R. Iga, H. Okamoto, Y. Miyamoto, and H. Takeuchi, *Electron. Lett.*, 37, 1, 2001.

61. H. Kawanishi, Y. Yamauchi, N. Mineo, Y. Shibuya, H. Murai, K. Yamada, and H. Wada, *Photon. Technol. Lett.*, 13, 954, 2001.

62. N. Mineo, K. Yamada, S. Sakai, and T. Ushikubo, *OECC 2000*, postdeadline paper 2-7.

63. K. Takagi, Y. Miyazaki, H. Tada, T. Aoyagi, T. Nishimura, and E. Omura, in *Proc. 2001 Int'l Conf. InP and Related Materials*, WA3-2, 2001, p. 432.

64. F. Devaux, S. Chelles, J. C. Harmand, N. Bouadma, F. Huet, M. Carre, and M. Foucher, in *Proc. IOOC95*, 1995, p. 56, paper FB3-2.

65. R. Weinmann, D. Baums, U. Cebulla, H. Haisch, D. Kaiser, E. Kuhn, E. Lack, K. Satzke, J. Weber, P. Wiedemann, and E. Zielinski, *Photon. Technol. Lett.*, 8, 891, 1996.

66. K. Wakita, K. Yoshino, I. Kotaka, S. Kondo, and Y. Noguchi, in *Proc. ECOC96*, Th. B.3.2, Brussels, Belgium, p. 1011.

67. B. N. Gomatam and N. G. Anderson, *IEEE J. Quantum Electron. Lett.*, 28, 1496, 1992.

68. S. Kuwano, N. Takachio, K. Iwashita, T. Otsuji, Y. Imai, and T. Enoki, K. Yoshino, and K. Wakita, in *Proc. OFC96*, 1996, paper PD25.

69. N. Shirai, H. Arimoto, K. Watanabe, A. Taike, K. Shinoda, J. Shimizu, H. Sato, T. Ido, T. Tsuchiya, M. Aoki, and S. Tsuji, ECOC 2002.

70. K. Wakita, Y. Toyoda, and T. Kojima, *Jpn. J. Appl. Phys.*, Vol. 41, Part 1, No. 2B, 2002, p. 1175.

71. K. Wakita, I. Kotaka, S. Matsumoto, R. Iga, S. Kondo, and Y. Noguchi, *Jpn. J. Appl. Phys.*, Vol. 37, Part 1, No. 3B, 1998, p. 1432.

72. S. Kondo, and Y. Noguchi, K. Tsuzuki, M. Yuda, S. Oku, *Jpn. J. Appl. Phys.*, Vol. 41, Part 1, No. 2B, 2002, p. 1171.

73. R. Sahara, K. Morito, K. Sato, Y. Kotaki, H. Soda, and N. Okazaki, *IEEE Photon. Technol. Lett.*, 7, 1004, 1995.

74. J. Shimizu, M. Aoki, T. Tsuchiya, M. Shirai, A. Taike, T. Ohtoshi, and S. Tsuji, *Electron. Lett.*, 38, 821, 2002.

75. Y. Miyazaki, H. Tada, S. Tokizaki, K. Takagi, Y. Hanamaki, T. Aoyagi, and Y. Mitsui, ECOC 2002, paper 10.5.6.

76. J. F. Fells, M. A. Gibbon, I. H. White, G. H. B. Thompson, R. V. Peny, C. J. Armistead, E. M. Kimber, J. D. Moule, and E. J. Thrush, *Electron. Lett.*, 30, 1168, 1994.

77. K. Yamada, N. Nakamura, Y. Matsui, T. Kunii, and Y. Ogawa, *IEEE Photon. Technol. Lett.*, 7, 1157, 1995.

78. K. Kawano, O. Mitomi, K. Wakita, I. Kotaka, and M. Naganuma, *IEEE J. Quantum Electron.*, QE-28, No. 1, pp. 224–230, 1992.

79. T. Ishibashi, N. Shimizu, H. Ito, T. Nagatsuma, and T. Furuta, Tech. Dig., UEO'97, 1997, p. 166.

80. M. Yoneyama, Y. Miyamoto, K. Hagimoto, N. Shimizu, T. Ishibashi, and K. Wakita, *Electron. Lett.*, 34, 1, 1998.

81. S. Kodama, T. Yoshimatsu, and H. Ito, *Electron. Lett.*, 38, 1575, 2002.

82. S. Kodama, T. Yoshimatsu, and H. Ito, Optical Fiber Communication Conference and Exhibit (OFC 2003), paper ThX5, Atlanta, Georgia, USA, March 23–28, 2003.

3 High-Speed LiNbO$_3$ Optical Modulators

O. Mitomi

CONTENTS

3.1 INTRODUCTION

The ferroelectric material LiNbO$_3$ (LN) has been extensively applied to optical devices. It has excellent electro-optic and optical properties, that is, it has a large electro-optic effect and is capable of a high-speed response. It is also transparent for infrared light and it is easy to fabricate into low-loss channel waveguides by diffusing titanium. Consequently, various high-performance optical waveguided

LiNbO$_3$ devices have been developed for the terminal functions of external intensity modulators, phase modulators, and multi/demultiplexers, as well as switch arrays for optical fiber network systems. In particular, these LiNbO$_3$ devices are very useful for optical wavelength-division-multiplexing (WDM) systems because of the possibility of operation in the range of wide-wavelength infrared light with a single device due to its transparence.

LiNbO$_3$ external modulators have been developed for extensive use in high-speed and long-distance optical fiber transmission systems. This is because they can offer the advantages of modulation exceeding 10 Gbits/s combined with a low driving voltage, and they can eliminate the dynamic laser wavelength chirping which limits the span-rate system product due to their fiber dispersion characteristics. LiNbO$_3$ external modulators can also offer pure phase modulation in coherent systems and can realize various optical signal processors. As the bit rate of optical network systems becomes higher, it becomes more difficult to drive a modulator with a high voltage due to the restrictions of electrical instruments, in particular, electrical driving amplifiers. Therefore, reduction of the driving voltage of an LN modulator with a broadband characteristic is an extremely important issue for realizing future high-speed optical transmission systems.

This chapter describes linear electro-optic Ti-diffused LiNbO$_3$ devices, particularly traveling-wave high-speed modulators, with respect to the device design and the fabrication procedures for optical network systems operating within the 1.3–1.6 μm wavelength region.

3.2 DEVICE DESIGN

3.2.1 OPERATION PRINCIPLES

The LiNbO$_3$ crystal demonstrates the linear electro-optic effect (Pockels effect) providing a change in refractive index proportional to the applied electric field E [1], that is,

$$\Delta n = -n^3 \cdot r \cdot E/2, \tag{3.1}$$

where n and r are the relevant refractive index and electro-optic coefficients, respectively, both based on crystal orientation. The largest electro-optic coefficient in LiNbO$_3$ is r_{33}, which is introduced when the polarized fields of the light and applied voltage are in the c-axis (z-axis) of the LiNbO$_3$ crystal. Using numerical values of $r_{33} \sim 30 \times 10^{-12}$ m/V and n_{33} $(= n_e)$ ~2.15 at a wavelength λ of 1.5 μm, an electric field of 10 V/μm is found to produce a refractive index change of about 1.5 $\times 10^{-3}$.

This index change induces an optical phase modulation of

$$\Delta \phi = k_o \cdot L \cdot \Delta n, \tag{3.2}$$

where $k_o = 2\pi/\lambda$ and L is the electrode length of the device. The optical phase modulation can be used to accomplish optical intensity modulation in several ways, namely, (1) interferometrically (Mach-Zehnder modulator, balanced bridge switch), and (2) phase match control (directional coupler). Figure 3.1 shows these Ti-diffused

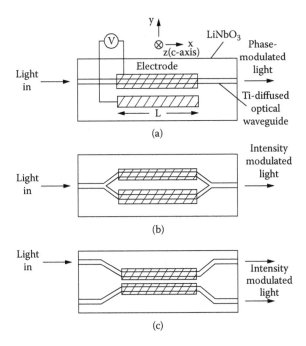

FIGURE 3.1 Scheme of typical waveguided z-cut LiNbO$_3$ devices. (a) phase modulator, (b) intensity modulator using Mach-Zehnder interferometer, and (c) intensity modulator using directional coupler.

single-mode channel waveguided device structures using z-cut LiNbO$_3$ crystals, where a low-refractive-index buffer layer of SiO$_2$ film is formed between the electrode and the LiNbO$_3$ substrate to prevent light absorption by the metal.

For external modulators, the absence of optical wavelength chirping in modulated signals is essential in high-speed and long-distance optical transmission systems. A Mach-Zehnder interferometric modulator has an ideal zero chirp [2] with push-pull *operation*, i.e., an increase in the refractive index of one waveguide causes a corresponding decrease in that of the other. Consequently, a Mach-Zehnder interferometer is normally utilized for high-speed intensity modulators, whereas a directional coupler is used for integrated switching devices.

3.2.2 DESIGN OF TI-DIFFUSED OPTICAL WAVEGUIDES

The LiNbO$_3$ optical waveguides are formed through thermal diffusion (at about 1000°C for over 10 hr) of titanium as will be described in a later section. The refractive index increase Δn_{Ti} is related to [3]

$$\Delta n_{Ti} = A \cdot C_{Ti}{}^a, \tag{3.3}$$

where A is a constant dependent on optical wavelengths, C_{Ti} is the titanium concentration in LiNbO$_3$, and a is a constant ($a = 0.8$ for extraordinary and 0.5 for ordinary indices).

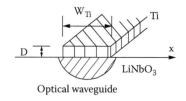

Optical waveguide

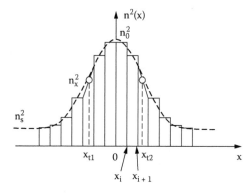

FIGURE 3.2 Approximate index profile of optical waveguide for numerical analysis.

From the diffusion equation, the Ti concentration distributions in LiNbO₃ can be expressed using Gaussian functions for the depth direction and error functions for the lateral direction of the LiNbO₃ substrate.

For analyzing Ti-diffused graded-index LiNbO₃ optical waveguides, as shown in Figure 3.2, many calculation methods have been extensively used, for example, the WKB method, the linear segment method (LSM) [4], and the modified step segment method (MSSM) [5]. On the other hand, computer numerical analyses by the finite element method (FEM) and the finite differential method (FDM) are useful for various waveguide structures. Figure 3.3 shows the calculated results for (a) the propagation constants of optical waveguides and (b) insertion losses of butt couplings between optical waveguides and single-mode fibers, where d_ℓ and d_d, respectively, are the lateral and depth diffusion lengths. The Ti-pattern width and thickness should be selected under the conditions of a single-mode optical waveguide and low-loss butt coupling with a fiber.

Low insertion losses in a LiNbO₃ device are of fundamental importance for optical network systems. Factors contributing to loss include waveguide propagation (both absorption and scattering), reflection, waveguide bending, and Y-branching as well as fiber-waveguide coupling. Propagation loss can be reduced using optimal fabrication conditions, as will be described in a later section. Reflection loss is eliminated using an antireflection coating. Bending loss has been reduced by determining optimal bend geometries using S-shaped or raised-cosine-shaped transitions for Ti-diffused waveguides. Conventional Y–branches have a relatively large loss of more than 0.5 dB. Low-loss branch waveguides have been developed, which use decreasing refractive indices near the branching region and raised-cosine-shaped waveguides to reduce radiation losses [5,6,7,8]. The Y-branches achieve a low loss of less than 0.2 dB.

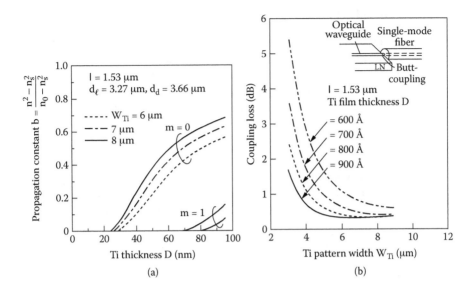

FIGURE 3.3 Calculated results for Ti-diffused LiNbO$_3$ optical waveguides. (a) normalized propagation constants of optical waveguide [5] and (b) coupling losses between optical waveguide and single-mode fiber [16].

3.2.3 ELECTRODE CONFIGURATION

The potential modulation bandwidths or high-speed properties of LiNbO$_3$ optical devices depend on the type of electrode and its microwave characteristics. Two types of electrodes, lumped and traveling-wave electrodes, have been used for waveguided LiNbO$_3$ devices. Figure 3.4(a) and Figure 3.4(b) illustrate the structures of the two types of devices. The lumped types are normally used in integrated optical devices such as switch matrices, and the traveling-wave types are used in high-speed modulators.

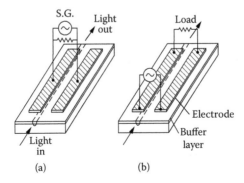

FIGURE 3.4 Electrode configurations for (a) lumped and (b) traveling-wave modulator.

3.2.3.1 Lumped-Type Devices

With the lumped devices shown in Figure 3.4(a), the electrodes act like a capacitor with its bandwidth limited by either the product of capacitance, C, and load resistance, R, or the transit time for a light wave to travel through the devices. Since the effect of capacitance is normally dominant for LiNbO$_3$ devices and limits the bandwidth, the modulation voltage, V_m, induced in the electrode becomes

$$V_m = V_S/(2 + j\omega_m \cdot C \cdot R), \tag{3.4}$$

where ω_m is the angular frequency of the microwave and V_S is the voltage of the signal generator having an inner impedance of R. The bandwidth Δf_L for which V_m is reduced by $1/\sqrt{2}$ (:electrical 3dB bandwidth) from its value when $\omega_m = 0$ then becomes

$$\Delta f_L \text{ (Hz)} = 1/(\pi \cdot R \cdot C) = 1/(\pi \cdot n_m^2 \cdot C_o \cdot R \cdot L), \tag{3.5}$$

where $C = n_m^2 \cdot C_o \cdot L$. C_o is the capacitance when all dielectrics are air, L is electrode length, and n_m is the effective refractive index of the microwave. That is, $n_m = (\varepsilon_{\text{eff}})^{1/2}$ where ε_{eff} is the effective RF relative dielectric constant.

Figure 3.5 shows the calculated capacities per unit length for symmetrical coplanar strips on a LiNbO$_3$ substrate, where $\varepsilon_{\text{eff}} = (1 + \varepsilon_S)/2$ without the use of a buffer layer between the electrode and LiNbO$_3$ substrate. ε_S is the dielectric constant for the LiNbO$_3$ substrate, and $\varepsilon_S = (\varepsilon_{x \text{ or } y} \cdot \varepsilon_z)^{1/2}$, $\varepsilon_{x, y} = 43$, and $\varepsilon_z = 28$. Then, $n_m = 4.2$. Also shown is the potential bandwidth-length product when R is 50 Ω. The width W_E must be sufficiently wide so that static electric resistance does not become the limiting factor in achieving bandwidth. Consequently, W_E is about 10 μm. On the other hand, the use

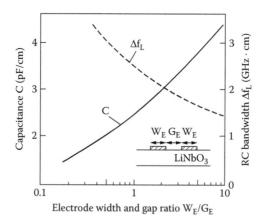

FIGURE 3.5 Calculated capacitance and RC bandwidth for symmetric strip electrodes on LiNbO$_3$ substrate for a lumped modulator assuming $R = 50$.

of a larger gap G_E can create a smaller capacitance. However, this increases the required driving voltage. As a result, G_E is normally fixed at about the width of an optical waveguide ($G_E = 5 - 10$ μm). Therefore, capacitance reaches about 2~3 pF/cm, and potential bandwidth Δ $f_L \cdot L$ would be ~ 2 GHz·cm for LiNbO₃ devices.

3.2.3.2 Traveling-Wave Type

3.2.3.2.1 Frequency Response of Modulators

With the traveling-wave device shown in Figure 3.4(b), the electrode forms an extension of the driving transmission line and should have the same characteristic impedance as the signal generator and cable. Both light and microwaves propagate in the same direction. The bandwidth is not limited by electrode capacitance but rather by the difference in velocity (velocity mismatch) between light and microwaves, and by the microwave propagation loss. The velocity mismatch depends on the LiNbO₃ material and the thickness of the buffer layer in addition to the electrode structures. Effective modulation voltage V_m induced through the electrode for light propagating in the waveguide, that is, the frequency response of the traveling-wave modulators, is then written as [9]

$$V_m(f) = \frac{1}{L}\int_0^L v_m(z,f)dz$$

$$= V_{m0} \cdot |(1+\rho_1) \cdot \exp(j\beta_o \cdot L) \cdot (V_+ + \rho_2 \cdot V_-)/[\exp(j\beta_e \cdot L) + \rho_1 \cdot \rho_2 \cdot \exp(-j\beta_e \cdot L)]|,$$

$$(3.6)$$

$$\rho_1 = (Z_m - Z_o)/(Z_m + Z_o),\ \rho_2 = (Z_\ell - Z_m)/(Z_\ell + Z_m),$$

$$\beta_o = \omega_m \cdot n_o/c,\ \beta_e = \omega_m \cdot n_m/c - j\alpha_m,$$

$$V\pm = \exp(\pm j\phi_\pm) \cdot \sin\phi_\pm /\phi_\pm,$$

$$\phi_\pm = (\beta_e \pm \beta_o) \cdot L/2,$$

$$V_{mo} = Vs \cdot Z_\ell/(Z_\ell + Z_o),\ \omega_m = 2\pi f,$$

where v_m is the effective modulation voltage of an electrical signal propagating through the electrodes at position z, c is the velocity of light in a vacuum, f the modulation frequency, Z_O the internal impedance of the signal generator, Z_ℓ the load resistance, Z_m the characteristic impedance of the electrode, n_m the microwave refractive index, n_O the optical refractive index, and α_m the microwave attenuation constant of the electrode, as shown in Figure 3.6.

Here, the response of Equation 3.6 shows the characteristic of small-signal ($V_{m0}/V_\pi \ll 1$) modulation. The modulators are, however, operated with large-signal modulation, such as pulse modulation, in optical transmission systems. Therefore, it is necessary to consider the large-signal characteristics for practical use [10].

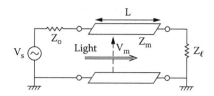

FIGURE 3.6 Microwave equivalent circuit for traveling-wave type modulator.

The microwave attenuation constant α_m consists of

$$\alpha_m = \alpha_c + \alpha_d + \alpha_r, \tag{3.7}$$

where α_c is the conductor loss induced by the skin effect of microwaves, α_d the dielectric loss, and α_r the radiation loss caused by the LN substrate resonance. α_d and α_r are negligibly small for below-50-GHz-band devices, as will be described in Section 3.6. The conductor loss α_c is

$$\alpha_c = \alpha_0 \cdot f^{1/2}, \tag{3.8}$$

where α_0 is a constant.

Normally, the characteristic impedance is equal to the load resistance ($Z_m = Z_o = Z_\ell = 50\ \Omega$). This is an impedance-matching condition. Then $\rho_1 = \rho_2 = 0$ and V_m is reduced to the familiar expression of

$$V_m = V_{mo} \cdot \exp(-j\ \phi_+) \cdot \sin\phi_+/\phi_+. \tag{3.9}$$

For previous LN modulators, the frequency responses were limited by a velocity mismatch. When the microwave attenuation constant α_m is assumed to be zero, the bandwidth (electrical 3 dB) obtained from Equation 3.9 is

$$\Delta f_{T1}(\text{Hz}) = 1.4c/(\pi\ |n_m - n_o|\ L\ ). \tag{3.10}$$

Here, the electrical 3-dB bandwidth Δf means that the frequency f of $V_m(f)$ becomes $V_m(f) = V_{m0}/\sqrt{2}$. It can be seen from Equation 3.10 that it is essential to reduce this mismatch in the inherent velocity to achieve high-speed devices. Since driving voltage, V, is found to be in inverse proportion to the electrode length L, established from Equation 3.2, V_π and Δf_{T1} are in a trade-off relationship. As a result, the term $\Delta f_{T1}/V_\pi$ becomes a positive indicator of device performance. Consequently, the potential bandwidth of the traveling-wave device had been restricted to about $\Delta f_{T1} \sim 9$ GHz·cm for previous LiNbO$_3$ devices of $n_m = 4.2$ and $n_o = 2.2$ with a wavelength of $\lambda = 1.55\ \mu$m. Therefore, in broader band operations, this type of electrode has considerable advantage over the lumped type.

If velocity matching could be achieved ($n_m = n_o$), the bandwidth should be restricted by the attenuation constant α_m. Normally, α_m approximately depends on the conductor loss α_c, that is, $\alpha_m \cong \alpha_c \propto f^{1/2}$. Then, the bandwidth becomes

$$\Delta f_{T2}\ (\text{GHz}) = 41/(\alpha_{c0} \cdot L), \tag{3.11}$$

where α_{c0} is the microwave power attenuation constant (dB/cm) at a frequency of $f = 1$ GHz and L is expressed in centimeters. α_{c0} is determined from the configurations of the electrode. In this case, the positive indicator of device performance is $\Delta f_{T2}/v_\pi^2$ obtained from Equation 3.11. The performance of high-speed LN modulators is, therefore, determined only by the product of driving voltage V and electrode length L, $V_\pi L$, and the conductor loss α_{c0} under both impedance and velocity matching conditions. Therefore, here, we introduce a parameter, p, defined as [11]

$$p = \alpha_{c0} \cdot V_\pi L, \qquad (3.12)$$

which can be calculated from an LN modulator structure and expresses a figure of merit of traveling-wave high-speed optical modulators. From Equation 3.11 and Equation 3.12, the driving voltage and 3-dB bandwidth of modulators are then expressed as

$$V_\pi \Big/ \sqrt{\Delta f_{T2}} = 0.156 \cdot p \qquad (3.13)$$

for Δf_{T2}(GHz) of the electrical 3-dB bandwidth. According to Equation 3.13, a smaller p can result in an excellent modulator, and V_π and Δf_{T2} can be determined from the electrode length L and the value of p.

3.2.3.2.2 Design of Device Structures

The coplanar waveguide (CPW) or the asymmetrical coplanar strip (A-CPS) is normally used for traveling-wave electrodes, as illustrated in Figure 3.7(a) and Figure 3.7(b).

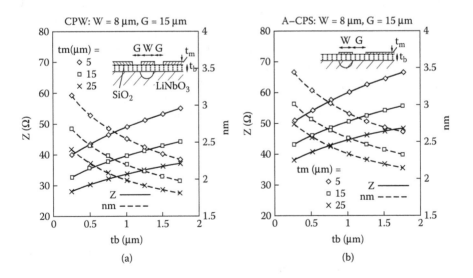

(a)   (b)

FIGURE 3.7 Calculated microwave effective index and characteristic impedance for (a) coplanar waveguide and (b) asymmetric coplanar strip line, where electrode width $W = 8$ μm and gap $G = 15$ μm.

This is because they have a relatively large optical confinement factor (the overlap integral between the applied electric field and the optical field) and possess both lower propagation loss in the microwave region and good coupling to an external coaxial cable.

For optical devices, a propagating microwave dominant-mode in traveling-wave electrodes is assumed to be purely TEM (transverse-electromagnetic mode) in nature due to the extremely narrow width W and gap G of the electrodes (normally $W, G = $ ~10 μm, being approximately the optical waveguide width) in comparison to microwave wavelength $\lambda_m (\approx 1 \text{ cm})$. Consequently, the frequency dependence of microwave characteristics for the dominant mode had been confirmed to be negligibly small for the CPW or A-CPS electrodes, by full-wave analyses [12]. The quasi-static (quasi-TEM) analyses can be used to calculate microwave characteristics of the traveling-wave electrodes. Here, the microwave field distributions are first calculated using Laplace's equation and an analysis method described later, and the capacitances of electrodes are then obtained using Gauss' theorem. The microwave effective refractive index n_m, characteristic impedance Z, and conductor loss α_c are expressed as [13,14]

$$n_m = \varepsilon_{eff}^{1/2} = (C/C_o)^{1/2} \tag{3.14}$$

$$Z_m = 1/[c \cdot (C \cdot C_o)^{1/2}] = Z_o/n_m \tag{3.15}$$

$$\alpha_c = (\varepsilon_o/\mu_o)^{1/2} \cdot Rs \cdot (\delta Z_o/\delta n)/(2Z_m), \tag{3.16}$$

where C is the capacitance per unit length, and C_o and Z_o are capacitance and characteristic impedance, respectively, when all dielectrics are air. Furthermore, μ_o and ε_o are the magnetic and dielectric permeabilities of free space, Rs is the skin resistance ($:Rs = (\pi \cdot \mu_o \cdot f \cdot \rho)^{1/2}$, ρ the electrode resistivity), and $\delta Z_o/\delta n$ denotes the derivative of Z_o with respect to incremental recession of the electrode surface.

The driving voltage (half-wavelength voltage) V_π means that the difference of optical phase modulation $\Delta\phi$ in Equation 3.2 becomes $\Delta\phi = \pi$, that is, $\Delta\phi = \Delta\phi_1 - \Delta\phi_2 = \pi$ for optical intensity modulation utilizing a Mach-Zehnder interferometer where $\Delta\phi_1$ and $\Delta\phi_2$ are the phase modulations in both Mach-Zehnder waveguide arms. From Equation 3.1 and Equation 3.2, $V_\pi L$ is approximately given by

$$V_\pi L = \lambda/(2\Delta n), \tag{3.17}$$

$$\Delta n = r_{33} n_e^3 \overline{E}_z/2,$$

where λ is the optical wavelength, and n_e and r_{33} are the relevant refractive index and electro-optic coefficient in the c-axis of the LiNbO$_3$ substrate, respectively, as mentioned for Equation 3.1. $\overline{E}_z$ is the effective microwave electrical field in the c-axis at the optical waveguides when the electrodes are biased with 1 V, and $\overline{E}_z$ is calculated from the overlap integrations between the microwave and optical wave fields.

The microwave field distributions and characteristics can be calculated using, for example, the conformal transformation method, the Fourier transform domain method (FTD), and the finite element method (FEM) [13]. The computer numerical analyses by the FEM and FDM are convenient for designing various device structures. However, the conformal transformation method can be used for designing simple structures of electrodes, in which both the electrode and buffer layer are sufficiently thin in comparison with the gap G of the electrode. Figure 3.7(a) and Figure 3.7(b) show examples calculated using the modified conformal transformation method [15] for coplanar waveguides and asymmetrical coplanar strips on LiNbO$_3$ substrates covered by a SiO$_2$ layer, as a function of the SiO$_2$ layer thickness t_b where the electrode thickness t_m is a parameter. As can be seen in Figure 3.7, the value of n_m decreases as t_b and t_m increase. Z increases as t_b increases, whereas it decreases as t_m increases. These relations can be deduced from Equation 3.14 and Equation 3.15.

3.3 DEVICE FABRICATION

Ti-diffused LiNbO$_3$ optical characteristics depend greatly on crystal and buffer-layer (SiO$_2$) qualities as well as fabrication parameters. In particular, Ti-diffusion conditions influence the propagation loss of optical waveguides, and buffer layer quality affects characteristic instability such as DC and thermal drifts.

With respect to these matters, a fabrication method was described for z-cut LiNbO$_3$ modulators [16,17]. The fabrication processes are shown in Figure 3.8.

3.3.1 OPTICAL WAVEGUIDE

Commercially available optical-grade LiNbO$_3$ crystals have been employed to fabricate the optical waveguides. The amount of Fe that causes optical damage proved to be less than 1 ppm, and the deviation in the refractive index of the wafer in terms of accuracy for mass production proved to be less than $\pm 1 \times 10^{-4}$. Additionally, the wafer employed was confirmed not to have any crucial defects through the use of x-ray topography.

The Ti-pattern is formed by a lift-off process and Ti-film is deposited using the electron beam evaporation method (Figure 3.8, parts (a)–(d)). Redeviation from the set value for a Ti-pattern formed on a wafer is less than $+0.1$ µm for Ti-film width and within $\pm 1.5\%$ for Ti-film thickness.

The wafers are contained in a quartz tube and diffusion is carried out at 1,000°C for over 10 hr in oxygen carrier gas with water vapor. To fabricate low-loss waveguides with good reproducibility, the degree of water vapor (H$_2$O) introduced into the carrier gas (O$_2$) should be optimized, since the introduction of water vapor into a diffusion atmosphere is known to greatly affect waveguide quality. Optical propagation loss was confirmed to be less than 0.2 dB/cm for the optimized diffusion conditions shown in Figure 3.9. Diffusion lengths in the depth direction, d$_d$, and the lateral direction, d_ℓ were about 3.6 µm and 3.3 µm, respectively.

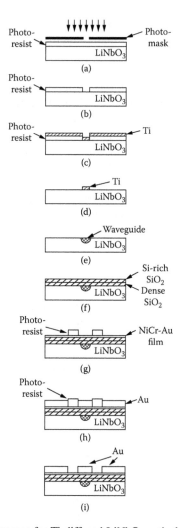

FIGURE 3.8 Fabrication process for Ti-diffused LiNbO₃ optical devices.

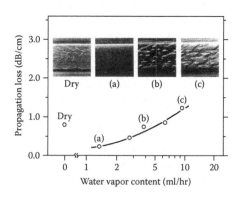

FIGURE 3.9 Effects of water vapor on propagation loss using oxygen as a carrier gas [17].

3.3.2 BUFFER LAYER AND ELECTRODE

Although Ti-diffused LiNbO$_3$ waveguide devices have characteristic instabilities due to DC and thermal drifts (characteristic changes due to environmental temperature changes), they can be eliminated by optimizing buffer layer properties. DC drift is considered to occur because of the movement of electric charges on the surface of the LiNbO$_3$ crystal or within the buffer layer when a DC bias is applied between the electrodes. Therefore, the effective electric field applied to the waveguides decreases. Hence, it is important to form a dense buffer layer [18]. Furthermore, thermal drift can be eliminated by forming a layer that quickly attenuates the electric charge because drift occurs due to the presence of a local electric charge that is generated by a pyroelectric effect between the electrodes [19]. Dual oxidized-silicon films were used for these purposes (Figure 3.8(f)). The first layer, the one nearest to the waveguide, is dense SiO$_2$, and the second layer is silicon-rich SiO$_2$ or amorphous Si for the purpose of eliminating pyroelectric charge. The dense SiO$_2$ and silicon-rich SiO$_2$ films were formed using the ECR plasma-CVD method, which enables deposition under relatively low temperature conditions.

To obtain thick electrodes, an electroplating method is employed to reduce conductor loss. That is, a NiCr-Au film is deposited onto the buffer layer and an electrode pattern is then photolithographically defined using a photoresist (Figure 3.8(g)). Subsequently, using the pattern as a guide, Au is grown by electroplating (Figure 3.8(h)). Finally, the NiCr-Au film deposited on the electrode gap is removed by ion milling (Figure 3.8(i)).

Figure 3.10 shows the temperature characteristics of driving point shift, that is thermal drift, for the fabricated Mach-Zehnder modulator employed with the dual-buffer layer structure. The dual-buffer layer clearly improves the device stability.

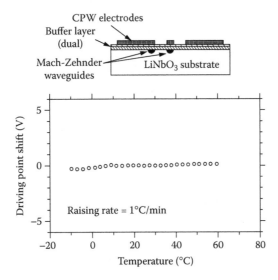

FIGURE 3.10 Temperature characteristics of driving point shift (thermal drift) for a fabricated LiNbO$_3$ device.

3.3.3 FIBER COUPLING AND PACKAGING

The various system applications of LiNbO₃ optical devices necessitate the development of practical techniques to permanently attach single-mode fibers. In general, the near-field distributions of Ti: LiNbO₃ channel waveguides adequately match the mode field of fibers so that butt-coupling techniques can be used between the LiNbO₃ waveguides and fibers. Figure 3.11 shows the butt-coupling and packaging techniques used for LiNbO₃ devices. A glass bead is used for reinforcement and is fixed with UV-cured epoxy to butt-coupled single fibers. Furthermore, to enable packages to be used over a wide range of temperatures, the fiber is buckled and fixed, taking into account the thermal distortion obtained from the differences in the thermal expansion coefficient between components in the packages. Anisotropically etched silicon or mechanically scraped glass V-grooves are used for the array coupling of fibers, as shown in Figure 3.11(c). The excess loss for the connection was less than 0.1 dB, and an average insertion loss of less than 0.3 dB could be achieved in these couplings.

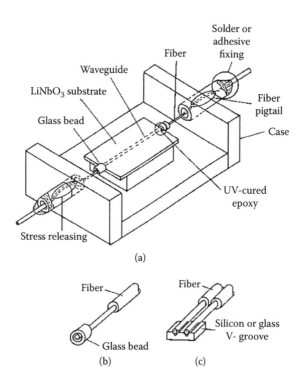

FIGURE 3.11 Techniques of fiber coupling to LiNbO₃ waveguides. (a) coupling and packaging of single-fiber pigtail, (b) single-fiber pigtail with glass bead, and (c) fiber array with silicon or glass V-groove.

3.4 BROADBAND OPTICAL MODULATORS

3.4.1 METHODS FOR VELOCITY MATCHING

The most important performance features of optical modulators are determined by the 3-dB bandwidth and driving voltage with a characteristic impedance system of 50 Ω, where the two factors are in a trade-off relationship, as indicated by Equation 3.10 and Equation 3.11. The main restriction hindering the improvement of broadband traveling-wave modulation is the velocity mismatch between light and microwaves, which is caused by the high dielectric constant of the LN substrate itself.

To overcome this problem, techniques of correcting velocity mismatching have been suggested, by which the effective index n_m (~4.2) for microwaves can be made to approach the effective index n_o (~2.2) for light. To reduce the velocity mismatch, n_m was decreased by using the lower dielectric constant of the SiO₂ buffer layer or/and air between the central conductor and grounding electrodes. As shown in Figure 3.12, these proposals have been divided into (a) methods that

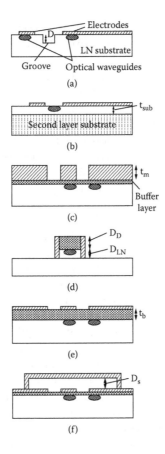

FIGURE 3.12 Sectional device configurations for reducing the velocity mismatch.

make use of the dielectric constant of air by providing grooves between waveguide channels [20], (b) methods that use thin LN on a second-layer substrate having a relatively low dielectric constant [21], (c) methods that use the dielectric constant of the air between the electrodes by increasing the thicknesses of both the central and grounding electrodes [22,23,24,25], (d) methods of accumulating low-dielectric-constant materials on a ridge-type optical waveguide by utilizing parallel-plate electrodes [26], (e) methods that utilize thicker buffer layer of low-dielectric-constant materials between the electrodes and optical waveguides [27], (f) methods that make use of the shielding plane laid over the electrodes and air gap [28], and (g) methods of pseudo-velocity matching that utilize phase-reversal electrode patterns as pseudorandom coded sequences [29].

3.4.2 DEVICE STRUCTURES AND CHARACTERISTICS

3.4.2.1 Planar z-cut LiNbO₃ Substrate

Figure 3.13 shows the schematic diagrams for a conventional broadband optical modulator with z-cut Ti-diffused LiNbO$_3$ Mach-Zehnder interferometric optical waveguides and coplanar waveguide (CPW) electrodes. To decrease the velocity mismatch, thick CPW electrodes were introduced in the modulators. The central-conductor width W was set to be almost the same as the optical spot size in the optical waveguides to obtain high modulation efficiency. Furthermore, a relatively thick SiO$_2$ buffer layer was employed to increase characteristic impedance and to decrease microwave conductor loss.

As shown in Figure 3.7, with constant central-conductor width W and gap G, the value of n_m decreases as t_b and t_m increase, and Z increases as t_b increases, whereas it decreases as t_m increases. Therefore, there is an optimal t_m at which velocity matching is attained for any W, G, and t_b. Figure 3.14 shows the characteristics calculated using a quasi-TEM FEM for a CPW planar LN modulator, in which the electrode thickness t_{mv} is the value resulting in a velocity match of $n_m = 2.2$, and the microwave characteristics of Z, α_0 at $f = 1$ GHz and $V_\pi L$ under the velocity-matching condition as a function of the buffer-layer thickness t_b, where W is a constant of 8 μm and G is a parameter [11]. The cases of $G = 15$ μm show a conventional modulator structure. A 50-Ω characteristic-impedance system and velocity-matching condition of planar modulators are shown to be realized with a buffer-layer thickness of about 1.7 μm and an electrode thickness of 10 μm for $G = 15$ μm.

One fabricated modulator consists of an optimized Ti-diffusion optical waveguide, as described in Section 3.3. To improve thermal stability, dual oxidized-silicon films were used for the buffer layer of the modulator. Characteristics of the fabricated Mach-Zehnder optical modulator for the 1.5-μm optical wavelength region have been normally obtained: electrical 3-dB bandwidth of 10–20 GHz and driving voltage $V\pi$ of 4–5 V with the gap $G = 15$ μm and length $L = 3–4$ cm of the CPW electrodes. The frequency response was proven to be totally limited by the microwave conductor loss of αc in Equation 3.11. Optical insertion loss, including fiber coupling losses, was less than 3 dB and an extinction ratio of −25 dB was typically obtained.

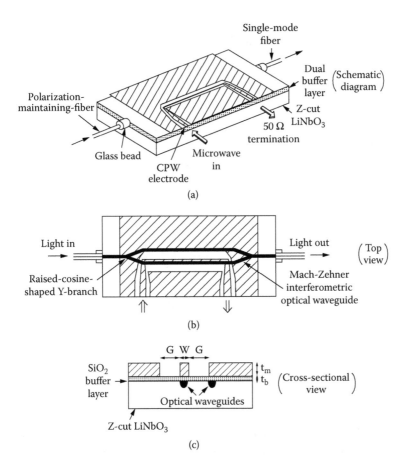

FIGURE 3.13 Device configurations of an intensity modulator consisting of a planar z-cut LiNbO$_3$ Mach-Zehnder interferometer and CPW traveling-wave electrode utilizing thicker SiO$_2$ buffer layer and electrodes.

Figure 3.15 illustrates another broadband optical modulator that uses the A-CPS electrodes, fabricated by increasing the thickness of both the center and grounding electrodes [22]. The fabricated modulator consists of a 1-μm-thick buffer layer, an electrode 10 μm thick and 2 cm long, and a mode-coupled Y-branch for the Mach-Zehnder optical waveguide. This had a 12-GHz electrical bandwidth with a driving voltage of 6.4 V, and a 2.2-dB insertion loss for a 1.54-μm optical wavelength. The bandwidth was limited by the velocity mismatch of $n_m = 2.7$ in Equation 3.10.

To achieve a velocity-matching condition, the above modulator structures must provide a characteristic impedance of the electrodes that is far lower than 50 Ω. This is because electrode capacitance increases markedly when the microwave effective-index decreases, and achieves a velocity-matching condition due to the use of thick

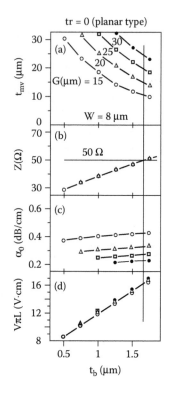

FIGURE 3.14 Calculated characteristics for CPW planar LN modulators, in which the electrode thickness t_{mv} is the value that results in a velocity match of $n_m = 2.2$, and the microwave characteristics of Z, α_0 at $f = 1$ GHz and $V_\pi L$ under the velocity matching condition as a function of the buffer-layer thickness t_b, where W is a constant of 8m and G is a parameter.

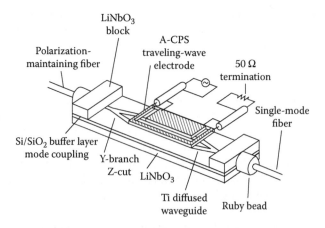

FIGURE 3.15 Schematic diagram of an intensity modulator consisting of a planar z-cut LiNbO$_3$ Mach-Zehnder interferometer and A-CPS traveling-wave electrode utilizing a relatively thick (1.2-m) SiO$_2$ buffer layer and relatively thick (10-m) electrode [22].

electrodes with a relatively thin buffer layer for attaining a relatively low driving voltage, as shown in Figure 3.14 for $G = 15$ μm. This impedance mismatch causes microwave reflections between the modulator and electrical instruments, and raises the serious problem of excess signal-bit error for high-speed modulations. Consequently, in order to create a broader bandwidth modulator, it is necessary to clarify the modulator structures to obtain both lower conductor loss and lower $V_\pi L$ under a condition of both perfect impedance and velocity matching.

From Figure 3.14, it is found that t_{mv} increases and α_0 decreases as G increases or t_b decreases for the planar LN substrate. On the other hand, at a constant G, Z and $V_\pi L$ decrease when t_b decreases. However, the dependence of Z on G is very small for any t_b. This is because, when G increases under the condition of velocity matching, the total capacitance $Ct (= Cs + Ce)$ of electrodes changes negligibly with a t_b due to the canceling of decreasing the LN-substrate capacitance Cs and increasing the capacitance Ce between surfaces of the thick center and earth electrodes up the buffer layer. That is, the characteristic impedance of electrodes is determined almost completely by the buffer-layer thickness under a velocity-matching condition with a proper thickness t_{mv} of electrodes. From Figure 3.14, a 50-Ω characteristic-impedance system of planar modulators is shown to be realized with a buffer-layer thickness of about 1.7 μm for any G. On the other hand, the $V_\pi L$ dependence on G is also seen to be weak in Figure 3.14(d). The weak dependence of $V_\pi L$ is due to the anisotropy of the LN dielectric constant. That is, the relative dielectric constant perpendicular (z-axis) to the substrate surface is small relative to that parallel to it, so that the perpendicular microwave field intensity Ez at the optical waveguides barely changes when G increases. As a result, higher performance planar modulators can be realized by employing a wider gap G [11].

3.4.2.2 Ridged z-cut LiNbO₃ Substrate

Modulator structures with a ridged LN substrate have been proposed for attaining a lower driving-voltage characteristic under a 50-Ω characteristic impedance system [24,30]. The schematic and cross-sectional configurations of ridged LN optical modulators having thicker CPW electrodes are shown in Figure 3.16. In this figure, the LN substrate is etched between the center and earth electrodes. The ridged structure can reduce the electrode capacitance and can achieve velocity matching with a relatively thin SiO₂ buffer layer. Furthermore, the structure has the merit of increasing the microwave field intensity at the Ti-diffused optical waveguides compared to the planar LN-substrate modulators. Figure 3.17 shows the product $V_\pi L$ as a function of the ridge depth for the optical wavelength of 1.55 μm where the CPW or A-CPS electrode width $W = 8$ μm and gap $G = 15$ μm. $V_\pi L$ decreases as the ridge depth increases for the buffer layer thicknesses of 0.75 and 1.20 μm, and becomes minimum at a depth between 3 and 4 μm. The decrease of $V_\pi L$ is caused by the concentration of the lines of electric force at the ridged optical-waveguide regions, which is due to the surrounding LiNbO₃ substrate being replaced by the lower dielectric-constant materials SiO₂ and air.

Figure 3.18 shows the calculated characteristics of ridged ($t_r = 5$ μm) LN modulators as a function of the buffer-layer thickness t_b. Z and $V_\pi L$ of the ridged

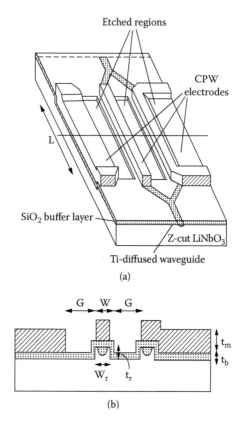

FIGURE 3.16 Schematic and cross-sectional configurations of ridged LN optical modulators having thicker CPW electrodes

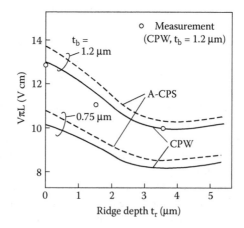

FIGURE 3.17 Product of the driving voltage V and the electrode interaction length L as a function of the ridge depth for the optical wavelength of 1.55 μm, where the electrode width $W = 8$ μm and gap $G = 15$ μm.

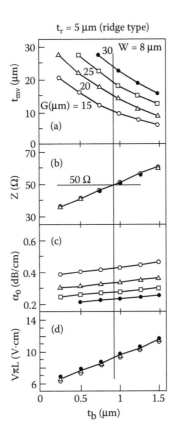

FIGURE 3.18 Calculated characteristics for CPW ridged LN modulators, in which the electrode thickness t_{mv} is the value resulting in a velocity match of $n_m = 2.2$, and the microwave characteristics of Z, α_0 at $f = 1$ GHz and $V_\pi L$ under the velocity matching condition as a function of the buffer-layer thickness t_b, where W is a constant of 8 μm and G is a parameter.

type are also almost completely determined by the buffer-layer thickness under a velocity-matching condition, and the 50-Ω characteristic impedance system can be achieved with a buffer-layer thickness of about 0.95 μm for the $t_r = 5$ μm ridged modulators.

On comparing the results for the planar and ridged modulators in Figure 3.14 and Figure 3.18, respectively, it is found that the required electrode thickness t_{mv} for the ridged type becomes thinner than that for the planar type at any G when both buffer-layer thicknesses t_b are the same. This is because the effective dielectric constant and electrode capacitance of LN substrate for the ridged type become lower than those for the planar type. $V_\pi L$ and Z of the ridged type are also smaller and greater, respectively, than those of the planar type under constant t_b. On the other hand, the conductor loss α_0 of the ridged type is the same as that of the planar type when both t_{mv} and G are the same. That is, α_0 is mainly determined by the electrode

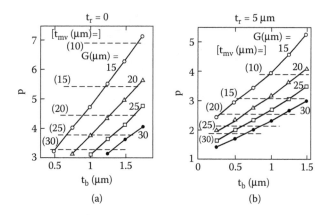

FIGURE 3.19 Calculated figures of merit p for CPW (a) planar ($t_r = 0$) and (b) ridged ($t_r = 5$ μm) modulators, respectively, under the velocity-matching condition.

thickness t_m. Under a characteristic impedance Z, the required buffer-layer thickness t_b of the ridge type becomes, however, thinner than that of the planar type, and as a result $V_\pi L$ of the ridged type is markedly smaller than that of the planar type.

Under the velocity-matching condition in Figure 3.14 and Figure 3.18, the calculated figures of merit p using Equation 3.12 are shown in Figure 3.19(a) and Figure 3.19(b) for planar ($t_r = 0$) and ridged ($t_r = 5$ μm) modulators, respectively. In the figures, the broken lines show the required thickness of t_{mv} shown in Figure 3.14(a) and Figure 3.18(a). p is known to become smaller when t_b decreases or G and t_{mv} increase, and t_{mv} provides a nearly constant p for any t_b and G in both planar and ridged types. This means that, in a lower characteristic-impedance system, thicker electrodes with a wider gap can be used to realize higher performance modulators due to the lower conductor loss α_0. Equation (3.13) and Figure 3.19(b) indicate that modulators having a wider gap and thicker electrodes with a ridged LN-substrate structure can realize a bandwidth and driving voltage of 40 GHz and far less than 3 V, or about 100 GHz and less than 5 V, as shown in Figure 3.20, under a 50-Ω characteristic-impedance system [11,31,32].

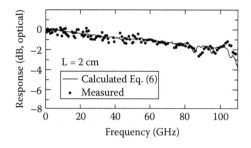

FIGURE 3.20 Optical response of fabricated ridged modulator [31].

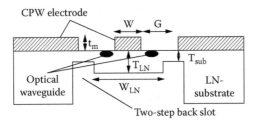

CPW electrode

FIGURE 3.21 Sectional view of a two-step back-slot structure for x-cut LiNbO$_3$ substrate modulator without a SiO$_2$ buffer layer.

3.4.2.3 Thinned x-cut LiNbO$_3$ Substrate

For the z-cut or x-cut modulators, a SiO$_2$ buffer layer is usually required for impedance matching. However, the buffer layer has been known to cause slight DC drift and thermal drift, and therefore, an auto-bias control circuit is needed to compensate the bias point drift for system operation. On the other hand, the conventional x-cut LN optical modulator is preferable, particularly for long-haul transmission systems, because the modulator has no wavelength chirping in the modulated output light due to the symmetrical device structure; this symmetry results in a perfect push-pull modulation for the two arms of the MZ waveguide. However, the existence of the buffer layer demands a higher driving voltage than that of z-cut LiNbO$_3$ modulators.

Figure 3.21 shows a two-step back-slot structure of an x-cut LiNbO$_3$ substrate without a SiO$_2$ buffer layer [33]. In this structure, the thickness, T_{LN}, of the substrate around the optical waveguides is increased to much larger than the optical mode size in order to avoid deformation and scattering. The thickness, T_{sub}, of the substrate near the ground electrodes is reduced in order to decrease the effective microwave index n_m and attain both the velocity-matching and impedance-matching conditions. Figure 3.22 shows the electrical S-parameters of electrodes and

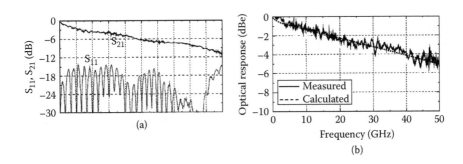

FIGURE 3.22 Electrical S-parameters of electrodes and the optical responses of the fabricated two-step back-slot modulator [33].

the optical responses of the fabricated modulators where V_π is 3 V, the 3-dB bandwidth is 25 GHz, and the fiber-to-fiber insertion loss is 5 dB in a structure of $T_{LN} = 15$ μm and $T_{sub} = 10$ μm. These modulators have sufficient capability for use in 40 Gbits/s optical fiber transmission systems [47].

3.5 FUNCTIONAL MODULATORS

Mach-Zehnder interferometric LN modulators are also indispensable for future radio communication systems as well as high-bit-rate dense-WDM systems because they can be operated over wide wavelength and frequency ranges. Consequently, LN modulators are applied to various optical signal processors such as optical subcarrier generators and frequency shifters.

3.5.1 PUSH-PULL TYPE

Mach-Zehnder modulators with two hot electrodes in the push-pull mode, as shown in Figure 3.23, can provide a chirpless signal and reduced driving voltage [34,35]. In addition, the push-pull modulators can generate an optical duobinary signal that allows a system's transmission bit rate at any given channel bandwidth to be increased; this is particularly important for dense WDM systems [36]. A push-pull modulator having a 3-dB bandwidth of 30 GHz and a driving voltage of less than 2 V in push-pull operation has been developed using a ridged z-cut LN substrate structure [37]. By utilizing wider gap $G = 50$ μm electrodes, a driver-less 40 Gbits/s z-cut LN modulator with a driving voltage of less than 1 V was also developed [38].

To control the chirp parameter in the z-cut LN modulator, a device structure in which a phase reversal electrode section is introduced in tandem with an inverted ferroelectric domain section in the LN crystal has been proposed, as shown in Figure 3.24 [39]. In the structure, the chirp parameter can be adjusted through the value of the ratio L_1/L_2 of the normal and inverted length, and becomes zero when $L_1 = L_2$.

3.5.2 MULTI-INPUT TYPE

Figure 3.25 shows an example of an optical signal-processing device, which is a single side-band suppression carrier (SSB-SC) signal generator consisting of four high-speed phase modulation waveguides [40]. The SSB-SC signal is generated by applying four individual electrical signals from the input ports. These devices are applicable in radio communication systems.

3.5.3 BAND-PASS TYPE

To attain high-speed (:millimeter wave frequency range) and low-power modulation, bandpass modulators have been proposed, which utilize phase synchronization between microwaves and light waves at a resonant frequency. These modulators can be accomplished by (1) periodic intermittent interaction electrodes [41], (2) resonant electrodes [42,43], or (3) periodic resonant electrodes [44].

Figure 3.26 shows the schema and frequency responses for a waveguided balanced-bridge Mach-Zehnder interferometric switch with a traveling-wave phase-reversed

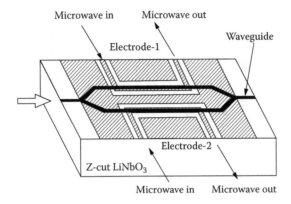

FIGURE 3.23 Schematic diagram of Mach-Zehnder modulators with two input ports.

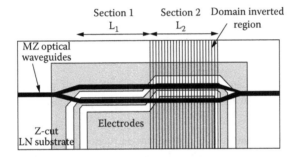

FIGURE 3.24 Device configuration of modulator introducing phase reversal electrodes for controlling chirp parameter in the modulated output light.

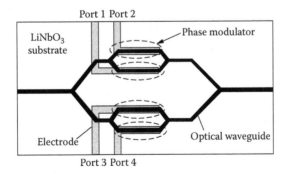

FIGURE 3.25 Device configuration of single side-band suppression carrier signal generator.

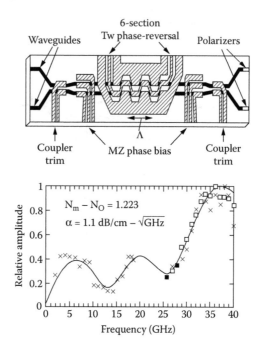

FIGURE 3.26 Schematic diagram and frequency response of a balanced-bridge Mach-Zehnder interferometric switch utilizing a traveling-wave phase-reversed electrode [41].

electrode [41]. The spatial period Λ of electrode phase reversals is $\Lambda = \lambda_{m0}/(n_m - n_o)$, where λ_{m0} is the electric wavelength at the bandpass center frequency. The device incorporates a phase-reversed coplanar waveguide having an $N = 6$ and $\Lambda/2 = 3.126$ mm section, and a microwave effective index of $n_m = 3.7$. The electrical power from the 50-Ω source was 407 mW at a bandpass center frequency of 36 GHz for full on-off modulation. This demonstrates the feasibility of optically multi/demultiplexing data rates of 72 Gbit/s.

3.6 CONSIDERATIONS FOR DESIGNING ULTRAHIGH-SPEED MODULATORS

Future optical network systems for the internet, cellular phones, and multimedia communication will require transmission technology exceeding 10 Tbits/s-band. New optical modulators will be expected to be capable of operation at bandwidths exceeding 100 GHz with a lower driving voltage. As the operating frequency increases to the millimeter-wavelength region, it is necessary to consider the key factors that limit high-speed modulation when designing device structures, that is, (a) the electrode waveguide dispersion, (b) the material dispersion, (c) the dielectric loss, and (d) the substrate resonance modes.

Figure 3.27 shows the measured frequency dependences of effective microwave index n_m and attenuation constant α in the electrodes for a ridged-substrate LiNbO$_3$

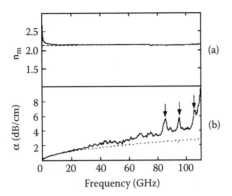

FIGURE 3.27 Measured frequency dependences of effective microwave index n_m and attenuation constant α in the electrodes for a ridged-substrate LiNbO₃ modulator [45].

modulator, where $W = 8$ μm, $G = 25$ μm, and $L = 2$ cm [45]. It is found from the figure that n_m is almost constant above 5 GHz, so that neither waveguide dispersion nor material dispersion is observed up to 100 GHz. On the other hand, α increases with frequency, and the conductor loss is found to be dominant in the frequency range of 1–20 GHz. α_{c0} at 1 GHz becomes 0.27 dB/cm, and the dashed line in Figure 3.27 indicates the calculated conductor loss. As the frequency increases above 20 GHz, the deviation between the calculated conductor loss and the measured α increases. This implies the occurrence of dielectric loss and radiation loss, which are marked by arrows in the figure. The dielectric loss α_d is due to the SiO₂ buffer-layer and LN substrate.

In the microwave attenuation constant α_m in Equation 3.7, the dielectric loss α_d can be calculated by treating the dielectric constants of the modulator materials as a complex coefficient in the microwave analysis. α_d is also obtained approximately as follows:

$$\alpha_d = \pi \cdot (\Gamma_{LN} \cdot \varepsilon_{LN} \cdot \tan \delta_{LN} + \Gamma_{SiO2} \cdot \varepsilon_{SiO2} \cdot \tan \delta_{SiO2}) \cdot f/(c \cdot n_m), \qquad (3.18)$$

where Γ, ε, and $\tan \delta$ are the confinement factor of microwave power, the relative dielectric constant, and the loss tangent of the LiNbO₃ substrate and the SiO₂ buffer layer, respectively. The confinement factor Γ is calculated with an electrical field by quasi-TEM analysis. $Tan\delta_{LN}$ and $\tan \delta_{SiO2}$ normally become about 0.004 and 0.01, respectively.

As the modulation frequency increases, the dependence of microwave propagation constant, that is, effective index n_m on modulation frequency, in addition to the higher order modes such as LN-substrate resonance modes, cannot be ignored. Therefore, full-wave vectorial analyses are, in general, required for designing a high-speed device, instead of the quasi-TEM analysis. Figure 3.28 shows the calculated results for the CPW planar or ridged LN modulators, using a full-wave analysis of vector FEM [46]. In the calculation, the surface of electrodes was treated as perfect

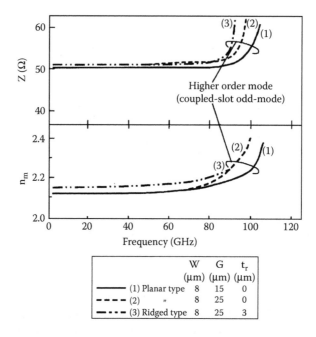

FIGURE 3.28 Calculated characteristic impedance and effective index dependence on modulation frequency, using a full-wave analysis.

electrical walls. It is shown in the figure that the CPW dominant modes are coupled with a higher order mode at around 100 GHz, and the characteristic impedance and the effective index increase significantly at that frequency. These higher-order modes may cause radiation loss or velocity mismatch in the devices. As a result, strict designs that take into consideration the factors limiting high-speed modulation will be required for ultrahigh-speed devices.

REFERENCES

1 I.P. Kaminow, *An Introduction to Electro-Optic Devices*, New York: Academic Press, 1974.
2 F. Koyama and K. Iga, "Frequency chirping in external modulators," *IEEE J. Lightwave Tech.*, LT-6, 87–93, 1988.
3 S. Fouchet, A. Carenco, C. Daguet, R. Guglielmi, and L. Riviere, "Wavelength dispersion of Ti induced refractive index change in LiNbO₃ as a function of diffusion parameters," *IEEE J. Lightwave Tech.*, LT-5 700–708, 1987.
4 H. Miyazawa and K. Kawano, "Three-dimensional analyses based on the linear segment method for Ti-diffused LiNbO₃ optical directional couplers," *Trans. IEICE Japan*, J71-C, 634–639, 1988.
5 T. Kitoh, K. Kawano, T. Nozawa, and H. Jumonji, "A study on design of high-speed and low-loss Ti:LiNbO₃ Mach-Zehnder optical modulator," *Trans. IEICE Japan*, J73-C-I, 332–339, 1990.

6   W. J. Minford, S.K. Korotky, and R.C. Alferness, "Low-loss Ti-LiNbO$_3$ waveguide bends at $\lambda = 1.3$ µm," *IEEE Tr. Microwave Theory and Tech.*, MTT-30, 1790–1794, 1982.

7   O. Hanaizumi, M. Miyagi, and S. Kawakami, "Wide Y-junctions with low losses in three-dimensional dielectric optical waveguides," *IEEE J. Quantum Electron.*, QE-21, 168–173, 1985.

8   M. Seino, T. Shiina, N. Mekada, and H. Nakajima, "Low-loss Mach-Zehnder modulator using mode coupling Y-branch waveguide," *Tech. Digest 13th ECOC' 87*, 113–116, 1987.

9   K. Kubota, J. Noda, and O. Mikami, "Traveling-wave optical modulator using a directional coupler LiNbO$_3$ waveguide," *IEEE J. Quantum Electron.*, QE-16, 754–760, 1980.

10   O. Mitomi, K. Noguchi, and H. Miyazawa, "Estimation of frequency response for high-speed LiNbO$_3$ optical modulators," *IEEE Proc. Optoelectron.*, 146, 99–104, 1999.

11   O. Mitomi, K. Noguchi, and H. Miyazawa, "Broadband and low driving-voltage LiNbO$_3$ optical modulators," *IEEE Proc. Optoelectron.*, 145, 360–364, 1998.

12   T. Kitazawa, D. Polifko, and H. Ogawa, "Analysis of CPW for LiNbO$_3$ optical modulator by extended spectral-domain approach," *IEEE Microwave Guid. Wave Lett.*, 2, 313–315, 1992.

13   K.C. Gupta, R. Garg, and I.J. Bahl, *Microstrip Lines and Slotlines*, Artech House, 1979.

14   M.V. Schneider, "Microstrip lines for microwave integrated circuits," *Bell Syst. Tech.*, 48, 1421–1444, 1969.

15   M.E. Davis, E.W. Williams, and A.C. Celestini, "Finite-boundary corrections to the coplanar waveguide analysis," *IEEE Trans. Microwave Theory and Tech.*, MTT-21, 594–596, 1973.

16   K. Kawano, T. Nozawa, M. Yanagibashi, and H. Jumonji, "Broad-band and low-driving-power LiNbO$_3$ external optical modulators," NTT Review, 1, 103–113, 1989.

17   T. Nozawa, K. Noguchi, H. Miyazawa, and K. Kawano, "Water vapor effects on optical characteristics in Ti:LiNbO3 channel waveguides," *Appl. Opt.*, 30, 1085–1089, 1991.

18   A.R. Beaumont, C.G. Atkins, and R.C. Booth, "Optically induced drift effects in lithium niobate electro-optic waveguide devices operating at a wavelength of 1.5 m," *Electron. Lett.*, 22, 1260–1261, 1986.

19   N. Mekada, M. Seino, T. Yamane, and H. Nakajima, "Thermally stabilized 1 × 4 Ti:LiNbO3 waveguide switch," *Technical Digest IGWO '89*, 4, 6–9, 1989.

20   H. Haga, M. Izutsu, and T. Sueta, "LiNbO$_3$ traveling-wave light modulator/switch with an etched groove," *IEEE J. Quantum Electron.*, QE-22, 902–906, 1986.

21   M. Izutsu, H. Haga, and T. Sueta,"Ultrafast Traveling-Wave Light Modulators with Reduced Velocity Mismatch," in *Picosecond Electronics and Optoelectronics*, G.A. Mourou, D.A. Bloom, and C.-H. Lee, Eds., Springer-Verlag, 1985, pp. 172–175.

22   M. Seino, N. Mekata, T. Yamane, and H. Nakajima, "12 GHz-bandwidth Ti:LiNbO$_3$ Mach-Zehnder modulator," Second Optoelectronics Conference, P.D. Tech. Digest, 1988, pp. 2–3.

23   G. K. Gopalakrishnan, C. H. Bulmer, W. K. Burns, R. W. McElahanon, and A. S. Greenblatt, "40 GHz, low half-wave voltage Ti:LiNbO3 intensity modulator," *Electron. Lett.*, 28, 826–827, 1992.

24   K. Noguchi, O. Mitomi, and H. Miyazawa, "A broadband Ti:LiNbO3 optical modulator with a ridge structure," *J. Lightwave Technol.*, 13, 1164–1168, 1995.

25 R. Madabhushi, "Microwave attenuation reduction techniques for wide-band Ti:LiNbO3 optical modulator," *Trans. IEICE, Japan, Trans. Electron.*, E81-C, 1321–131327, 1998.

26 K. Miura, M. Minakata, and S. Kawakami, "A broad-band traveling-wave optical modulator with high efficiency," Tech. Group OQE 87-26, IECE Japan, 1987, pp. 95–102.

27 K. Kawano, T. Kitoh, O. Mitomi, T. Nozawa, and H. Jumonji, "Wide-band and low-driving-power phase modulator employing a Ti:LiNbO3 optical waveguide at 1.5 μm wavelength," *IEEE Photon. Tech. Lett.*, T1, 3334, 1989.

28 K. Kawano, T. Kitoh, H. Jumonji, T. Nozawa, and M. Yanagibashi, "New traveling-wave electrode Mach-Zehnder optical modulator with 20 GHz bandwidth and 4.7 V driving voltage at 1.52 m wavelength," *Electron. Lett.*, 25, 1382–1383, 1989.

29 D.W. Dolfi, M. Nazarathy, and R.L. Jungerman, "40 GHz electro-optic modulator with 7.5 V drive voltage," *Electron. Lett.*, 24, 528–529, 1988.

30 K. Noguchi, O. Mitomi, K. Kawano, and H. Miyazawa, "Highly efficient 40-GHz bandwidth Ti:LiNbO3 optical modulator employing ridge structure," *IEEE Photon. Tech. Lett.*, 5, 52–54, 1993.

31 K. Noguchi, O. Mitomi, and H. Miyazawa, "Millimeter-wave Ti:LiNbO3 optical modulators," *J. Lightwave Technol.*, 16, 615–619, 1998.

32 K.Noguchi, H.Miyazawa, and O.Mitomi, "40-Gbit/s Ti:LiNbO3 optical modulator with two-stage electrode," Trans. IEICE, Japan, Trans. Electron., Vol. E81-C, no. 8, pp. 1316–1320, 1998.

33 J. Kondo, A. Kondo, K. Aoki, M. Imaeda, T. Mori, U. Mizuno, S. Takatsuji, Y. Kozuka, O. Mitomi, and M. Minakata, "40Gb/s X-cut LiNbO$_3$ optical modulator with two-step back-slot structure", *J. Lightwave Technol.*, 20, 2110–2114, 2002.

34 T. Namiki, N. Mekata, H. Hamano, T. Yamane, M. Seino, and H. Nakajima, "Low-driving-voltage Ti:LiNbO3 Mach-Zehnder modulator using a coupled line," in *Tech. Digest, Optical Fiber Communication Conference*, TUH4, 34, 1990.

35 S. K. Korotky, J. J. Veselka, C. T. Kemmerer, W. J. Minford, D. T. Moser, J. E. Watson, C. A. Mattoe, and P. L. Stoddard, "High-speed, low power optical modulator with adjustable chirp parameter," in *Tech. Digest, Conference of Integrated Photonics Research*, TuG2, 53–54, 1991.

36 K. Yonenaga, S. Kuwano, S. Norimatsu, and N. Shibata, "Optical duobinary transmission system with no receiver sensitivity degradation," *Electron. Lett.*, 31, 302–304, 1995.

37 K. Noguchi, O. Mitomi, and H. Miyazawa, "Push-pull type ridged Ti:LiNbO3 optical modulator," IEICE, Japan, Trans. Electron., E79-C, 27–31, 1996.

38 M. Sugiyama, M. Doi, S. Taniguchi, T. Nakazawa, and H. Onaka, "Driver-less 40 Gb/s LiNbO3 modulator with sub-1 V drive voltage," in *Tech. Digest, Optical Fiber Communication Conference*, PD FB6-2, 2002.

39 N. Courjal, H. Porte, A. Martinez, and J.-P. Goedgebuer, "LiNbO3 Mach-Zehnder modulator with chirp adjusted by ferroelectric domain inversion," *IEEE Photon. Tech. Lett.*, 14, 1509–1511, 2002.

40 S. Shimotsu, S. Oikawa, T. Saitou, N. Mitsugi, , K. Kubodera, T. Kawanishi, and M. Izutsu, "Single side-band modulation performance of a LiNbO$_3$ integrated modulator consisting of four-phase modulator waveguides," *IEEE Photon. Tech. Lett.*, 13, 364–366, 2001.

41 S. K. Korotky and J.J. Veselka, "Efficient switching in a 72-Gbit/s Ti:LiNbO$_3$ binary multiplexer/demultiplexer," *Tech. Digest OFC' 90*, TUH2, 1990.

42 M. Izutsu et al., "Millimeter-wave light modulator using LiNbO$_3$ waveguide with resonant electrode," *Technical Digest CLEO' 88*, PD14, 485–486, 1988.

43 S. Oikawa, T. Kawanishi, K. Higuma, Y. Matsuo, and M. Izutsu, "Double-stub structure for resonant-type optical modulators using 20-m-thick electrode," *IEEE Photon. Tech. Lett.*, 15, 221–223, 2003.

44 J.H. Schaffner and R.R. Hayes, "Ka-band integrated optic modulators with periodic intermittent interaction electrodes," *Integrated Photonics Research '91*, TuD9, 1991.

45 K. Noguchi, H. Miyazawa, and O. Mitomi, "Frequency-dependent propagation characteristics of coplanar waveguide electrode on 100 GHz Ti:LiNbO3 optical modulator," *Electron. Lett.*, 34, 661–662, 1998.

46 M. S. Alam, M. Koshiba, K. Hirayama, and Y. Hayashi, "Analysis of lossy planar transmission lines by using a vector finite element method," *IEEE Trans. Microwave Theory Tech.*, 43, 2466–2470, 1995.

47. J. Kondo, K. Aoki, A. kondo, T. Eiiri, Y. Iwata, A. Hamajima, T. Mori, Y. Mizuno, M. Imaeda, Y. Kozuka, O. Mitomi, and M. Minakata, "High-speed and low-driving-voltage thin-sheet X-cut LiNbO$_3$ modulator with Laminated low-dielectric-constant adhesive," *IEEE Photon. Tech. Lett.*, 17, 10, 2077–2079, 2005.

4 III-V Compound Semiconductor Electro-Optic Modulators

Nadir Dagli

CONTENTS

4.1 INTRODUCTION

The optical modulator is a key component for photonics. Optical fiber communications, microwave photonics, instrumentation, and optical signals processing all require optical modulators. Several different technology platforms can be used for the realization of optical modulators. Of these, $LiNbO_3$-based ferroelectric electro-optic modulators provide the most mature technology. Electro-optic polymers and compound semiconductors are also attractive technologies for optical modulators.

Compound semiconductors have unique properties that offer significant potential for high-performance modulators. These modulators can be grouped into two major types. The first is the electro-absorption type. These modulators take advantage of strong, electric field tunable absorption in quantum wells enabling very small devices. Such devices are covered in detail in chapter 2. The other type is the electro-optical modulator, which is the topic of this chapter.

Electro-optic modulators take advantage of the electro-optic properties of compound semiconductors. These materials do not have inversion symmetry and posses a linear electro-optic coefficient; in contrast, elemental semiconductors such as silicon have a center of symmetry and therefore possess no electro-optical coefficient. The linear electro-optic coefficients of compound semiconductors are 15 to 20 times smaller than that of LiNbO$_3$. However, this disadvantage can be overcome using other properties. One such property is the high refractive index of compound semiconductors. Compound semiconductor indices of refraction are around 3.5, compared to 2.2 for LiNbO$_3$. The index change experienced due to the linear electro-optic effect is proportional to the cube of the index of refraction. This property combined with the high index of refraction provides a fourfold improvement. Therefore, the index change for the same electric field is only about 5 times more efficient in LiNbO$_3$ compared to compound semiconductors. However, for the same applied voltage, a much higher electric field can be used in compound semiconductors. This (perhaps not so obvious) property is due to the dispersion of the dielectric constant. In ferroelectric materials such as LiNbO$_3$ there is a very strong ionic contributing to the dielectric constant which is dominant at microwave and millimeter-wave frequencies. But this contribution disappears at infrared and optical frequencies, and the relative dielectric constant decreases from about 35 to 4.5. On the other hand, the dielectric constant of compound semiconductors decreases from about 13 to 10 in the range from DC to optical frequencies.

Most electro-optic modulators take advantage of the traveling wave idea to reduce the drive voltage while keeping the bandwidth very high. This requires matching the velocity of the microwave signal on the modulator electrode with the optical signal in the optical waveguide. The microwave signal typically sees both the air and the electro-optic material, and its effective dielectric constant is about the arithmetic mean of the air and material dielectric constants. But the optical signal is almost entirely in the electro-optic material. As a result, in LiNbO$_3$, the optical signal travels faster than the electrical signal. This situation is just the opposite in compound semiconductors, and the microwave signal needs to be slowed for velocity matching. This can be done very effectively by increasing the electrode capacitance, which in turn can be done by reducing the electrode gap. Hence, high electric fields that overlap very well with the optical mode can be applied. This increases the efficiency of the modulation significantly. But the opposite has to be done in LiNbO$_3$; hence electric field strength cannot be increased to very high levels. Taking advantage of this observation along with advanced semiconductor processing techniques, it becomes possible to realize compound semiconductor modulators with better drive voltage and bandwidth characteristics than LiNbO$_3$ modulators.

Furthermore, the electro-optic efficiency of compound semiconductors can be enhanced significantly using quadratic electro-optic effects. This has led to the production of InP-based electro-optic modulators with 2.3-V drive voltages and 42-GHz bandwidth that are only about 3 mm long. A LiNbO$_3$ modulator with similar characteristics would have to be about 5 cm long. In addition, compound semiconductor modulators have the potential of integration with optical sources, detectors, and passive components that are usually also fabricated in compound semiconductors.

On the other hand, compound semiconductor waveguides have higher propagation loss than their $LiNbO_3$ counterparts. Moreover, the high index of refraction that helps to improve the electro-optic efficiency forces the optical mode to be well confined and therefore small. For this reason, mode transformers are usually needed to reduce fiber coupling losses.

This chapter starts with the description of physical effects used in III-V compound semiconductors. First, the linear electro-optic effect and the quadratic electro-optic effect are described. Formulas to predict the index of refraction and the linear and quadratic electro-optic coefficients along with experimental values are provided. This is followed by the description of the other effects, namely the plasma and band filling effects that can give index changes. Next, the different types of phase modulators used in compound semiconductors are introduced. Formulas are given to predict the drive voltage of a Mach-Zehnder type amplitude modulator when both linear and quadratic electro-optic effects are present. Traveling-wave electro-optic modulator design guidelines for high speed modulator applications follow this section. Next, specific high speed modulator examples are given. Finally a summary is presented.

4.2 PHYSICAL EFFECTS USED IN III-V COMPOUND SEMICONDUCTOR ELECTRO-OPTIC MODULATORS

4.2.1 LINEAR ELECTRO-OPTIC EFFECT

An applied electric field $E(V/m)$ to a material polarizes the atoms, molecules, or ions in the material by exerting oppositely directed forces on the positively and negatively charged parts of the atoms, ions, or molecules. This polarization creates an additional polarization field P which can be expressed $P = \chi E$, where χ is a numerical coefficient known as the dielectric susceptibility. In the presence of polarization the total electrical flux density $D(C/m^2)$ becomes $D = \varepsilon_0 E + \varepsilon_0 P = \varepsilon_0 (1 + \chi)$ $E = \varepsilon_0 \varepsilon_r E = \varepsilon E$, where $\varepsilon_0 = 8.5 \times 10^{-12} (F/m)$ is the dielectric permittivity of the vacuum, $\varepsilon_r = (1 + \chi)$ is the relative dielectric constant, and ε is the dielectric constant. The index of refraction of the crystal n is defined as $n = \sqrt{\varepsilon_r}$. Using this definition, $D = \varepsilon_0 n^2 E$. In general, in a dielectric material E and electric flux density D are related through the dielectric tensor.

$$\begin{bmatrix} D_x \\ D_y \\ D_z \end{bmatrix} = \begin{bmatrix} \varepsilon_{xx} & \varepsilon_{xy} & \varepsilon_{xz} \\ \varepsilon_{yx} & \varepsilon_{yy} & \varepsilon_{yz} \\ \varepsilon_{zx} & \varepsilon_{zy} & \varepsilon_{zz} \end{bmatrix} \begin{bmatrix} E_x \\ E_y \\ E_z \end{bmatrix} = \varepsilon_0 \begin{bmatrix} n_{xx}^2 & n_{xy}^2 & n_{xz}^2 \\ n_{yx}^2 & n_{yy}^2 & n_{yz}^2 \\ n_{zx}^2 & n_{zy}^2 & n_{zz}^2 \end{bmatrix} \begin{bmatrix} E_x \\ E_y \\ E_z \end{bmatrix}, \qquad (4.1)$$

where ε_{xx}, ε_{xy}, ... are constants of the medium. Using energy conservation arguments, it can be shown that the dielectric tensor should be symmetric, i.e., $\varepsilon_{kl} = \varepsilon_{lk}$ [1]. Therefore, the dielectric tensor has only six independent components. The dielectric tensor can always be diagonalized by using an appropriate coordinate transformation.

The coordinate axes in which the dielectric tensor is diagonal are known as the principle axes. One can rewrite Equation 4.1 as

$$
\begin{bmatrix} E_x \\ E_y \\ E_z \end{bmatrix} = \frac{1}{\varepsilon_0} \begin{bmatrix} \left(\dfrac{1}{n^2}\right)_{xx} & \left(\dfrac{1}{n^2}\right)_{xy} & \left(\dfrac{1}{n^2}\right)_{xz} \\ \left(\dfrac{1}{n^2}\right)_{xy} & \left(\dfrac{1}{n^2}\right)_{yy} & \left(\dfrac{1}{n^2}\right)_{yz} \\ \left(\dfrac{1}{n^2}\right)_{xz} & \left(\dfrac{1}{n^2}\right)_{yz} & \left(\dfrac{1}{n^2}\right)_{zz} \end{bmatrix} \begin{bmatrix} D_x \\ D_y \\ D_z \end{bmatrix} \tag{4.2}
$$

The electric energy density W_e in a medium is given as $W_e = \dfrac{1}{2} E \cdot D$. Using the relationship shown in Equation 4.2, this can be written as

$$
2\varepsilon_0 W_e = \left(\frac{1}{n^2}\right)_{xx} D_x^2 + \left(\frac{1}{n^2}\right)_{yy} D_y^2 + \left(\frac{1}{n^2}\right)_{zz} D_z^2 + 2\left(\frac{1}{n^2}\right)_{yz} D_y D_z + 2\left(\frac{1}{n^2}\right)_{xz}
$$
$$
D_x D_z + 2\left(\frac{1}{n^2}\right)_{xy} D_x D_y \tag{4.3}
$$

It should be noted that in Equation 4.3,

$$
\left(\frac{1}{n^2}\right)_{xx} \neq \left(\frac{1}{n_{xx}^2}\right)
$$

unless the principle axes are used as the coordinate system. Since the left-hand side of Equation 4.3 is a constant, the following association can be made:

$$
x = \frac{D_x}{\sqrt{C}}, \quad y = \frac{D_y}{\sqrt{C}}, \quad \text{and} \quad z = \frac{D_z}{\sqrt{C}}, \quad \text{where} \quad C = \sqrt{2\varepsilon_0 W_e}.
$$

With this association and using the abbreviated notation in which $xx \equiv 1$, $yy \equiv 2$, $zz \equiv 1$, $yz \equiv 4$, $xz \equiv 5$, and $xy \equiv 6$, Equation 4.3 can be rewritten as

$$
\left(\frac{1}{n^2}\right)_1 x^2 + \left(\frac{1}{n^2}\right)_2 y^2 + \left(\frac{1}{n^2}\right)_3 z^2 + 2\left(\frac{1}{n^2}\right)_4 yz + 2\left(\frac{1}{n^2}\right)_5 xz + 2\left(\frac{1}{n^2}\right)_6 xy = 1. \tag{4.4}
$$

In this equation, x, y, and z relate to the polarization of the optical field. For example, $y = z = 0$ signifies an x-polarized optical field. Equation 4.4 represents an ellipse and is known as the index ellipsoid or optical indicatrix. Using this equation, one can find the phase velocity of an optical wave propagating in the crystal in any arbitrary direction [2].

The refractive index of a material changes in response to an applied electric field [1]. The refractive index change can be expressed as a power series expansion in terms of the applied electric field as

$$\Delta\left(\frac{1}{n^2}\right)_{ij} = r_{ijk}E_k + R_{ijkl}E_kE_l + \cdots \tag{4.5}$$

E_k is the kth component of the applied electric field. The first term describes the linear dependence of the refractive index on the electric field and r_{ijk} is known as the linear electro-optic coefficient. The second term describes the quadratic dependence on the electric field and R_{ijkl} is known as the quadratic electro-optic coefficient. This effect is described in more detail in the next section.

Using the abbreviated notation defined earlier, the linear electro-optic effect can be described explicitly as

$$\begin{pmatrix} \Delta\left(\frac{1}{n^2}\right)_1 \\ \Delta\left(\frac{1}{n^2}\right)_2 \\ \Delta\left(\frac{1}{n^2}\right)_3 \\ \Delta\left(\frac{1}{n^2}\right)_4 \\ \Delta\left(\frac{1}{n^2}\right)_5 \\ \Delta\left(\frac{1}{n^2}\right)_6 \end{pmatrix} = \begin{pmatrix} r_{11} & r_{12} & r_{13} \\ r_{21} & r_{22} & r_{23} \\ r_{31} & r_{32} & r_{33} \\ r_{41} & r_{42} & r_{43} \\ r_{51} & r_{52} & r_{53} \\ r_{61} & r_{62} & r_{63} \end{pmatrix} \begin{pmatrix} E_x \\ E_y \\ E_z \end{pmatrix}. \tag{4.6}$$

With these changes, the index ellipsoid is written as

$$\left[\left(\frac{1}{n^2}\right)_1 + \Delta\left(\frac{1}{n^2}\right)_1\right]x^2 + \left[\left(\frac{1}{n^2}\right)_2 + \Delta\left(\frac{1}{n^2}\right)_2\right]y^2 + \left[\left(\frac{1}{n^2}\right)_3 + \Delta\left(\frac{1}{n^2}\right)_3\right]z^2$$

$$+ 2\left[\left(\frac{1}{n^2}\right)_4 + \Delta\left(\frac{1}{n^2}\right)_4\right]yz + 2\left[\left(\frac{1}{n^2}\right)_5 + \Delta\left(\frac{1}{n^2}\right)_5\right]xz + 2\left[\left(\frac{1}{n^2}\right)_6 + \Delta\left(\frac{1}{n^2}\right)_6\right]xy = 1 \tag{4.7}$$

The same expression can also be written as

$$(x\ y\ z)\begin{bmatrix}\left(\left(\dfrac{1}{n^2}\right)_1+\Delta\left(\dfrac{1}{n^2}\right)_1\right) & \left(\left(\dfrac{1}{n^2}\right)_6+\Delta\left(\dfrac{1}{n^2}\right)_6\right) & \left(\left(\dfrac{1}{n^2}\right)_5+\Delta\left(\dfrac{1}{n^2}\right)_5\right) \\ \left(\left(\dfrac{1}{n^2}\right)_6+\Delta\left(\dfrac{1}{n^2}\right)_6\right) & \left(\left(\dfrac{1}{n^2}\right)_2+\Delta\left(\dfrac{1}{n^2}\right)_2\right) & \left(\left(\dfrac{1}{n^2}\right)_4+\Delta\left(\dfrac{1}{n^2}\right)_4\right) \\ \left(\left(\dfrac{1}{n^2}\right)_5+\Delta\left(\dfrac{1}{n^2}\right)_5\right) & \left(\left(\dfrac{1}{n^2}\right)_4+\Delta\left(\dfrac{1}{n^2}\right)_4\right) & \left(\left(\dfrac{1}{n^2}\right)_3+\Delta\left(\dfrac{1}{n^2}\right)_3\right)\end{bmatrix}\begin{pmatrix}x\\y\\z\end{pmatrix}=1$$

$$(4.8)$$

In single-crystal material, crystal symmetry reduces the number of independent coefficients. For crystals having a center of symmetry, such as silicon, all r_{ijk} coefficients are zero; hence, such materials do not possess a linear electro-optic coefficient. III-V compound semiconductors do not have a center of symmetry. They have Zinc Blende or $\overline{4}3$ m crystal structure, which reduces the linear electro-optic tensor to

$$\begin{pmatrix}\Delta\left(\dfrac{1}{n^2}\right)_1\\[2mm]\Delta\left(\dfrac{1}{n^2}\right)_2\\[2mm]\Delta\left(\dfrac{1}{n^2}\right)_3\\[2mm]\Delta\left(\dfrac{1}{n^2}\right)_4\\[2mm]\Delta\left(\dfrac{1}{n^2}\right)_5\\[2mm]\Delta\left(\dfrac{1}{n^2}\right)_6\end{pmatrix}=\begin{bmatrix}0 & 0 & 0\\0 & 0 & 0\\0 & 0 & 0\\r_{41} & 0 & 0\\0 & r_{41} & 0\\0 & 0 & r_{41}\end{bmatrix}\begin{pmatrix}E_x\\E_y\\E_z\end{pmatrix}$$

$$(4.9)$$

In this case, there is only one linear electro-optic coefficient, r_{41}. Therefore, in the presence of an applied external field, using Equation 4.8 and Equation 4.9, the index ellipsoid can be expressed as

$$(x\ y\ z)\begin{bmatrix}\dfrac{1}{n^2} & r_{41}E_z & r_{41}E_y\\[2mm]r_{41}E_z & \dfrac{1}{n^2} & r_{41}E_x\\[2mm]r_{41}E_y & r_{41}E_x & \dfrac{1}{n^2}\end{bmatrix}\begin{pmatrix}x\\y\\z\end{pmatrix}=1$$

$$(4.10)$$

The spectral and compositional dependence of the electro-optic coefficient for different materials has been investigated experimentally and theoretically.

III-V compound semiconductors also possess piezoelectric and photoelastic properties. An applied electric field can contribute to an index change through a piezoelectrically induced photoelastic effect. In other words, the applied electric field can create strain in the material through the piezoelectric effect. The resulting elastic deformation can induce an index change through the photoelastic effect. Hence, r_{41} can be written as a combination of two parts:

$$r_{41} = r_{41}^s + r_{41}^p .$$ (4.11)

r_{41}^p is due to photoelastic contribution through the piezoelectric effect. This effect can only be observed at low frequencies and is also known as the false electro-optic effect [3]. r_{41}^s is the electro-optic coefficient observed at high frequencies well above the acoustic resonances of the material. It is also known as the clamped value measured at constant strain. Under this condition there is no elastic deformation, and r_{41}^s is the only contribution to r_{41}.

Table 4.1 summarizes the experimental reports of r_{41} for different III-V compound semiconductors at different wavelengths. In cases when a source reports both clamped and unclamped r_{41} values, only the clamped value relevant for high-frequency modulation is listed.

Several models have been developed to describe the spectral dependence of the linear electro-optic coefficient [15,16]. The model in [15] is based on tight binding calculations of the electronic susceptibility. The r_{41} predicted by this model shows a weak dispersion between 0.888 and 1.52 μm but a clear decrease at longer wavelengths [9]. The longer wavelength prediction is not observed in the experimental data. The model in [16] uses empirical parameters to describe the spectral dependence of the bandgap and the real part of the dielectric constant using the formula

$$r_{41}^s(\omega) = \left| \frac{1}{\varepsilon_1^2(\omega)} [E_0 g(\chi) + F_0] \right|$$ (4.12)

where $\varepsilon_1(\omega)$ is the frequency dependent dielectric constant given as

$$\varepsilon_1(\omega) = A_0 \left\{ f(\chi) + \frac{1}{2} \left[\frac{E_{g0}}{E_{g0} + \Delta_{s0}} \right]^{\frac{3}{2}} f(\chi_{s0}) \right\} + B_0$$ (4.13)

TABLE 4.1
Linear Electro-Optic Coefficients of In$_{1-x}$Ga$_x$As$_y$P$_{1-y}$

Material	(x, y)	Wavelength(μm)	r_{41} (10^{-12} m/V)	Source
GaAs	(1,1)	0.877	1.86 ± 0.2	[4]
GaAs	(1,1)	0.888	1.74 ± 0.2	[4]
GaAs	(1,1)	1.06	1.8 ± 0.08	[5]
GaAs	(1,1)	1.064	1.33	[6]
GaAs	(1,1)	1.09	1.72 ± 0.08	[7]
GaAs	(1,1)	1.15	1.68 ± 0.08	[7]
GaAs	(1,1)	1.15	1.58 ± 0.06	[8]
GaAs	(1,1)	1.15	1.63 ± 0.076	[7]
GaAs	(1,1)	1.3	1.6	[5]
GaAs	(1,1)	1.32	1.54 ± 0.08	[9]
GaAs	(1,1)	1.52	1.5 ± 0.08	[9]
GaAs	(1,1)	3.39	1.44 ± 0.04	[10]
GaAs	(1,1)	3.39	1.6 ± 0.3	[11]
GaAs	(1,1)	3.39	1.5 ± 0.15	[12]
GaAs	(1,1)	10.6	1.71 ± 0.05	[10]
GaAs	(1,1)	10.6	1.6 ± 0.3	[11]
GaAs	(1,1)	10.6	1.6 ± 0.3	[13]
InP	(0,0)	1.064	1.34	[6]
InP	(0,0)	1.208	1.54	[6]
InP	(0,0)	1.306	1.59	[6]
InP	(0,0)	1.5	1.68	[6]
InGaAsP	(0.16, 0.34) E_g = 1.122 eV	1.25	1.43	[14]
InGaAsP	(0.16, 0.34) E_g = 1.122 eV	1.32	1.44	[14]
InGaAsP	(0.1, 0.23) E_g = 1.193 eV	1.32	1.34	[14]

with

$$f(\chi) = \chi^{-2} \left[2 - (1+\chi)^{\frac{1}{2}} - (1-\chi)^{\frac{1}{2}} \right], \tag{4.14}$$

$$\chi = \frac{\omega}{E_{g0}} \quad \text{and} \quad \chi_{s0} = \frac{\omega}{(E_{g0} + \Delta_{s0})}.$$

$g(\chi)$ is given as

$$g(\chi) = \chi^{-2} \left[2 - (1+\chi)^{-\frac{1}{2}} - (1-\chi)^{-\frac{1}{2}} \right]. \tag{4.15}$$

TABLE 4.2
Parameters Used in the Calculation of $\epsilon_1(\omega)$ and $r_{41}^s(\omega)$

Material	A_0	B_0	E_0 (10^{-12} m/v)	F_0 (10^{-12} m/v)
InP	8.40	6.60	−42.06	91.32
GaP	22.28	0.92	−83.31	16.60
GaAs	9.29	7.86	−71.48	123.16
InAs	4.36	10.52	−30.23	197.88

The parameters used in the calculations for binary compound semiconductors are given in Table 4.2 [16].

For $In_{1-x}Ga_xAs_yP_{1-y}$ lattice matched to InP,

$$x = \frac{0.1894y}{(0.4184 - 0.013y)},$$

$$E_{g0} = 1.35 - 0.72y + 0.12y^2, \tag{4.16}$$

$$\Delta_{s0} = 0.12 + 0.3y + 0.11y^2.$$

For quaternary alloys A_0, B_0, E_0, and F_0, parameters can be interpolated to the desired alloy composition using the well-known formula

$$Q_{InGaAsP}(x, y) = (1-x)(1-y)B_{InP} + x(1-y)B_{GaP} + xyB_{GaAs} + (1-x)yB_{InAs} \tag{4.17}$$

where $Q(x, y)$ is the interpolated quaternary parameter and B is the appropriate binary parameter. This model predicts a flat spectral dependence for r_{41} for longer wavelengths and its results agree well with the experimental data. However, it also predicts a divergent behavior near the wavelengths corresponding to the bandgap energy, which is not supported by the experimental data. But this model could be useful to predict r_{41} at wavelengths away from the bandgap for different compositions for which experimental data are not available.

The most commonly used crystal orientation is [001]. In this case, if the modulating field is z directed, $E_x = E_y = 0$, and Equation 4.10 reduces to

$$(x \; y \; z) \begin{bmatrix} \dfrac{1}{n^2} & r_{41}E_z & 0 \\ r_{41}E_z & \dfrac{1}{n^2} & 0 \\ 0 & 0 & \dfrac{1}{n^2} \end{bmatrix} \begin{pmatrix} x \\ y \\ z \end{pmatrix} = 1 \tag{4.18}$$

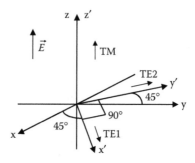

FIGURE 4.1 The axes of the index ellipsoid for an III-V compound semiconductor in the case of a z-directed external electric field. The dielectric tensor is diagonal in x, y, z, and x′,y′,z′ coordinate systems in the absence and the presence of the electric field, respectively. TE1, TE2, and TM show the direction of the main optical field of the two orthogonal TE and TM modes, respectively.

The index ellipsoid is no longer diagonal. It can be diagonalized using a simple coordinate transformation in the xy plane as shown in Figure 4.1. The new $x'y'$ axes are simply rotated 45° with respect to the xy axes. In the new $x'y'z$ coordinate system, the index ellipsoid can be written as

$$(x' \ y' \ z)\begin{bmatrix} \dfrac{1}{n^2} - r_{41}E_z & 0 & 0 \\ 0 & \dfrac{1}{n^2} + r_{41}E_z & 0 \\ 0 & 0 & \dfrac{1}{n^2} \end{bmatrix}\begin{pmatrix} x' \\ y' \\ z \end{pmatrix} =$$

$$(x' \ y' \ z)\begin{bmatrix} \dfrac{1}{(n+\Delta n_{x'})^2} & 0 & 0 \\ 0 & \dfrac{1}{(n+\Delta n_{y'})^2} & 0 \\ 0 & 0 & \dfrac{1}{n^2} \end{bmatrix}\begin{pmatrix} x' \\ y' \\ z \end{pmatrix} = 1$$

(4.19)

In this case there is no index change for an optical wave polarized in the z direction. The index of refraction for an optical wave polarized in the x' direction and propagating in the y' direction can be found using Equation 4.19 as

$$\frac{1}{(n+\Delta n_{x'})^2} = \frac{1}{n^2} - r_{41}E_z \quad \text{or} \quad (n+\Delta n_{x'}) = \frac{n}{\sqrt{1 - r_{41}n^2 E_z}}$$

Since $r_{41}n^2 E_z \ll 1$, $(n + \Delta n_{x'}) \approx n\left(1 + \frac{1}{2}r_{41}n^2 E_z\right)$,

$$\Delta n_{x'} = \frac{1}{2}r_{41}n^3 E_z. \tag{4.20}$$

Similarly, the index of refraction for an optical wave polarized in the y' direction and propagating in the x' direction can be found as

$$\Delta n_{y'} = -\frac{1}{2}r_{41}n^3 E_z. \tag{4.21}$$

In practice, the z axis is along the 001 crystal plane, and the x' and y' axes are along the 110 and 1$\overline{1}$0 crystal axis. Therefore, in such material a vertically applied electric field in the 001 direction increases the index of refraction by $\Delta n_{x'}$ in the 110 direction and decreases in by $\Delta n_{y'}$ in the 1$\overline{1}$0 direction. In other words, the index increases along one of the two mutually orthogonal directions parallel to the surface of the crystal, and it decreases by the same amount in the other orthogonal direction. These directions correspond to the cleavage planes of 001-oriented material. No index change is observed in the 001 direction. This implies that an electric field vertically applied to an 001-oriented crystal will only modulate the TE mode of the optical waveguide in which the main electric field component of the optical mode is either in 110 or 1$\overline{1}$0 directions; in other words, it is tangential to the surface. No modulation will result for the TM mode, which has its main electric field component in 001 direction, i.e., normal to the surface of the crystal.

4.2.2 QUADRATIC ELECTRO-OPTIC EFFECT OR ELECTRO-REFRACTIVE EFFECT

The refractive index of all materials shows a quadratic dependence on an applied external electric field as given in Equation 4.5. This effect exists in all materials and is known as the Kerr effect. In single-crystal material the number of independent coefficients is reduced due to crystal symmetry. Using the abbreviated notation defined earlier for Zinc Blende or $\overline{4}$3m crystal structure, the quadratic electro-optic tensor can be expressed as

$$
\begin{pmatrix}
\Delta\left(\frac{1}{n^2}\right)_1 \\
\Delta\left(\frac{1}{n^2}\right)_2 \\
\Delta\left(\frac{1}{n^2}\right)_3 \\
\Delta\left(\frac{1}{n^2}\right)_4 \\
\Delta\left(\frac{1}{n^2}\right)_5 \\
\Delta\left(\frac{1}{n^2}\right)_6
\end{pmatrix}
=
\begin{bmatrix}
R_{11} & R_{12} & R_{12} & 0 & 0 & 0 \\
R_{12} & R_{11} & R_{12} & 0 & 0 & 0 \\
R_{12} & R_{12} & R_{11} & 0 & 0 & 0 \\
0 & 0 & 0 & R_{44} & 0 & 0 \\
0 & 0 & 0 & 0 & R_{44} & 0 \\
0 & 0 & 0 & 0 & 0 & R_{44}
\end{bmatrix}
\begin{pmatrix}
E_x^2 \\
E_y^2 \\
E_z^2 \\
E_y E_z \\
E_x E_z \\
E_x E_y
\end{pmatrix}
\tag{4.22}
$$

The R_{ij} coefficients are usually very weak since they describe a second-order perturbation to the index of refraction. However in compound semiconductors, other physical effects can strongly enhance these coefficients at wavelengths near the bandgap energy of the material. These effects are the changes in the absorption near the bandgap due to applied electric fields. In bulk material, energy bands are tilted in the presence of an electric field. This tilting increases the penetration of the electron and hole wave functions into the bandgap. This in turn makes it is possible to excite an electron from the valence band to the conduction band with a photon of energy less than the energy of the bandgap. As a result, absorption increases with increasing external field and an absorption tail extending to shorter energies or longer wavelengths develops. This phenomenon is known as the Franz-Keldysh effect [17,18]. Absorption can be described using the imaginary part of a complex optical index. The real and imaginary parts of the complex index of refraction are related through the Kramers-Kronig relation. Therefore, a change in absorption generates a change in the refractive index. This effect also has a quadratic dependence on the applied electric field and shows strong dispersion near the bandgap of the material, since absorption changes are near the bandgap. This effect is also known as the Kerr effect or electro-refractive effect. It should be noted that the index change obtained this way comes with an absorption change, and sometimes compromises have to be made.

Table 4.3 gives the experimental reports of quadratic electro-optic coefficients for different materials at different wavelengths. GaAs is well characterized, but data for other materials are somewhat scarce.

TABLE 4.3
Reported Measured Quadratic Electro-Optic Coefficients of $In_{1-x}Ga_xAs_yP_{1-y}$

Material	(x,y)	Wavelength (μm)	R_{11} (10^{-21} m²/V²)	R_{12} (10^{-21} m²/V²)	$R_{11} \times R_{12}$ (10^{-21} m²/V²)	Source
GaAs	(1,1)	1.09	29±25%			[7]
GaAs	(1,1)	1.15	20±25%			[7]
GaAs	(1,1)	1.09		24±25%		[7]
GaAs	(1,1)	1.15		17±25%		[7]
GaAs	(1,1)	1.06			43±5	[5]
GaAs	(1,1)	1.3			13±5	[5]
GaAs	(1,1)	1.32	9.3±2.8			[9]
GaAs	(1,1)	1.52	3.2±2.3			[9]
GaAs	(1,1)	1.32		5.1±1.9		[9]
GaAs	(1,1)	1.52		5.1±2.6		[9]
GaAs	(1,1)	1.09			54	[19]
GaAs	(1,1)	1.15			40	[19]
InGaAsP	(0.16, 0.34), E_g = 1.122 eV	1.25			58	[14]
InGaAsP	(0.16, 0.34), E_g = 1.122 eV	1.32			31	[14]
InGaAsP	(0.1, 0.23), E_g = 1.193 eV	1.32			8	[14]

As argued before, the most commonly used crystal orientation is [001]. In this case if the modulating field is z directed, $E_x = E_y = 0$, and Equation 4.22 reduces to

$$\left[\Delta\left(\frac{1}{n^2}\right)_1 \Delta\left(\frac{1}{n^2}\right)_2 \Delta\left(\frac{1}{n^2}\right)_3 \Delta\left(\frac{1}{n^2}\right)_4 \Delta\left(\frac{1}{n^2}\right)_5 \Delta\left(\frac{1}{n^2}\right)_6\right]^T = [R_{12}E_z^2 \; R_{12}E_z^2 \; R_{11}E_z^2 \; 0 \; 0 \; 0]^T$$

With given perturbation and using the form of the index ellipsoid given in Equation 4.8, we can express the index ellipsoid as

$$(x \; y \; z)\begin{bmatrix} \frac{1}{n^2}+R_{12}E_z^2 & 0 & 0 \\ 0 & \frac{1}{n^2}+R_{12}E_z^2 & 0 \\ 0 & 0 & \frac{1}{n^2}+R_{11}E_z^2 \end{bmatrix}\begin{pmatrix} x \\ y \\ z \end{pmatrix} = 1 \qquad (4.23)$$

To express the index ellipsoid in the (x', y', z') coordinate system introduced in Figure 4.1, we need to transform Equation 4.23. However, this transformation leaves the equation unchanged. In other words, the unprimed coordinates in Equation 4.23 could be replaced by primed coordinates. Then, using an analysis similar to starting with Equation 4.19, we obtain the index changes TE and TM polarizations experience as

$$\Delta n_{x'} = \Delta n_{TE} = \frac{1}{2}R_{12}n^3E_z^2 \qquad (4.24)$$

and

$$\Delta n_{y'} = \Delta n_{TM} = \frac{1}{2}R_{11}n^3E_z^2. \qquad (4.25)$$

Usually $R_{11} \approx R_{12}$ and most of the literature quotes a single quadratic electro-optic coefficient $R = R_{11} \approx R_{12}$. The R coefficient is a strong function of wavelength of operation or the detuning from the band edge.

The calculated changes in the index of refraction and the absorption coefficient as a function of photon energy for GaAs and $In_{0.76}Ga_{0.24}As_{0.52}P_{0.48}$ at different electric field strengths are shown in Figure 4.2.

The index change increases rapidly towards the bandgap. The peak of the increase shifts away from the band edge with increasing electric field. For wavelengths very close to the bandgap index, the change decreases and it should eventually

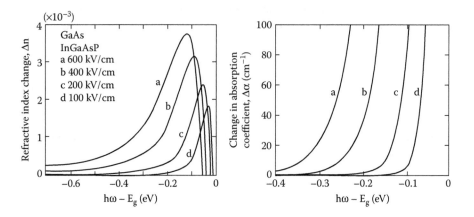

FIGURE 4.2 The calculated change in the index of refraction and absorption coefficient as a function of photon energy for GaAs and $In_{0.76}Ga_{0.24}As_{0.52}P_{0.48}$ at different electric field strengths [22].

be negative. Figure 4.4 shows the change in the refractive index in $In_{0.76}Ga_{0.24}As_{0.52}P_{0.48}$ as a function of the applied electric field at two different wavelengths. For photon energies far below the bandgap, the refractive index change for a given electric field can be expressed as

$$\Delta n = k(\lambda)E^2. \tag{4.26}$$

There is a quadratic dependence on the electric field and the wavelength dependent coefficient $k(\lambda)$ can be expressed for GaAs as

$$k(\lambda) = 3.45 \times 10^{-16} \exp\left(\frac{3}{\lambda^3}\right) cm^2\ V^{-2} \tag{4.27}$$

where wavelength is expressed in micrometers [22]. Figure 4.3 shows the spectral variation of the quadratic electro-optic coefficient for GaAs based on Equation 4.27 along with the values given in Table 4.3. The values based on the formula seem to be lower at shorter wavelengths. Therefore, the formula can be used as a worst-case estimate. The index change due to the quadratic electro-optic effect for quaternary material was also calculated in [22]. It is reported that for $In_{0.76}Ga_{0.24}As_{0.52}P_{0.48}$, $k(1.55\ \mu m) = 5.8 \times 10^{-15}\ cm^2\ V^{-2}$[22]. Another analytical formula is given for InGaAsP material system as [21]

$$R(\Delta E) = 150 \times 10^{-21} \exp(-8.85|\Delta E|)\ m^2/\ V^2, \tag{4.28}$$

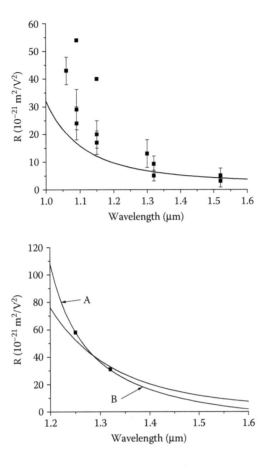

FIGURE 4.3 (a) Variation of the quadratic electro-optic coefficient for GaAs based on Equation 4.27. Experimental values given in Table 4.3 are also plotted. (b) Variation of the quadratic electro-optic coefficient for $In_{0.84}Ga_{0.16}As_{0.34}P_{0.66}$. Curves A and B are based on Equation 4.29 and Equation 4.28, respectively. Experimental values given in Table 4.3 are also plotted.

where ΔE is the energy difference in eV between the photon energy and the bandgap of the material. The electrorefractions in bulk GaAs, InP, GaSb, InSb, and InGaAsP were also calculated [22,23]. In cases where experimental data are not available, the value and the dispersion of the quadratic electro-optic coefficient expression can be estimated using the expressions given in [24] and [22]. The expression given in [24] predicts

$$R_{11} \approx R_{12} = \left| \frac{1}{\varepsilon_1^2(\omega)} [G_0 h(\chi) + H_0] \right|, \qquad (4.29)$$

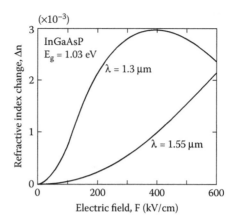

FIGURE 4.4   The change in the refractive index Δn in $In_{0.76}Ga_{0.24}As_{0.52}P_{0.48}$ as a function of the applied electric field at two different wavelengths [22].

where $\varepsilon_1(\omega)$ is the dielectric function defined earlier in Equation 4.13. $h(\chi)$ is given as

$$h(\chi) = \chi^{-2}\left[2 - (1+\chi)^{-\frac{3}{2}} - (1-\chi)^{-\frac{3}{2}}\right],\qquad(4.30)$$

with the earlier definition of χ given in Equation 4.14. Although the constants G_0 and H_0 can be calculated, they are typically found by curve fitting the experimental data. For $In_{0.84}Ga_{0.16}As_{0.34}P_{0.66}$, these constants are given as $G_0 = 3.43 \times 10^{-19}$ m^2/V^2 and $H_0 = 2.85 \times 10^{-18}$ m^2/V^2. For a given InGaAsP composition, once the G_0 and H_0 are determined, the spectral shape of the quadratic electro-optic coefficients can be calculated.

At high electric fields, it is possible to raise the refractive index by about 3×10^{-3} at photon energies near the bandgap. However, this index change is accompanied by a large absorption change due to the Franz-Keldysh effect. In order to get predominantly an electro-optic effect absorption, change should be kept to a minimum. The relative variation of the real and imaginary parts of the index of refraction with applied field is quantified using the figure of merit $\Delta n/\Delta k$, where Δk is the change in the imaginary part of the refractive index. The terms Δk and $\Delta \alpha$ are related as $\Delta \alpha = 4\pi\Delta k/\lambda$.

Absorption can be improved significantly in quantum well (QW) materials due to strong excitonic effects. The spectra of such strong excitonic absorption are very sharp and are localized in the vicinity of wavelengths corresponding to the bandgap of the QW. It is possible to strongly modify the absorption in a QW close to the band edge with applied electric fields, which is known as the Quantum Confined Stark Effect (QCSE). This strong change in the absorption also creates an accompanying index change. Figure 4.5 [25] shows Δn and $\Delta n/\Delta k$ as a function of the electric field for two

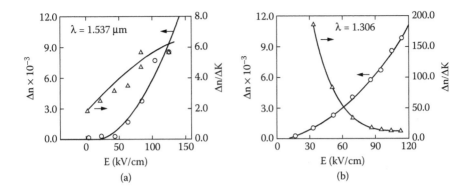

FIGURE 4.5 Change in the refractive index Δn and the ratio of the changes in the real and imaginary parts of the refractive index $\Delta n/\Delta k$ as function of the electric field for two MQW samples. (a) Ten periods of 85 Å InP barriers and 85 Å InGaAsP wells of bandgap energy 1.57 μm at 1.537 μm. (b) Five periods of 250 Å InP barriers and 70 Å InGaAsP wells of bandgap energy 1.33 μm at 1.16 μm [25].

MQW samples at two different wavelengths. In both cases, Δn shows a quadratic dependence on the applied field. $R(1.537\ \mu m) = 6.73 \times 10^{-13}\ cm^2\ V^{-2}$ for an 85-A InGaAsP QW of bandgap energy 1.57 μm within 85-A InP barriers. $R(1.306\ \mu m) = 7.32 \times 10^{-13}\ cm^2\ V^{-2}$ for a 70-A InGaAsP QW of bandgap energy 1.33 μm within 250-A InP barriers [25]. These values are more than two orders of magnitude larger than the corresponding bulk material values. But this does not necessarily make a much better modulator since the overlap of the optical mode with the QW material in a typical modulator is also a few percent. However, the spectral width of the absorption in a QW material is much narrower compared to that of bulk material. Furthermore, QCSE red-shifts this relatively narrow resonance, so that for a given wavelength detuning, the value of $\Delta n/\Delta k$ is larger in the QW material than in bulk material.

4.2.3 PLASMA EFFECT

The plasma effect is the accompanying index change associated with the free carrier absorption. For n-type GaAs of carrier concentration N, the index change Δn_N with respect to undoped material is

$$\Delta n_N = -9.6 \times 10^{-21} \frac{N}{n\mathcal{E}^2},$$

where n is the index of refraction and $\mathcal{E}$ is the photon energy. For p-type GaAs of carrier concentration P, the corresponding index changes for intra- and inter-band transitions are

$$\Delta n_{PInter} = -6.3 \times 10^{-22} \frac{P}{\mathcal{E}^2}.$$

and

$$\Delta n_{PIntra} = -1.8 \times 10^{-21} \frac{P}{n\mathcal{E}^2}.$$

As these expressions indicate, removal of electrons or holes in a doped semi-conductor reduces free carrier absorption, which in turn increases the index of refraction.

4.2.4 BAND-FILLING EFFECT

In doped material, the position of the Fermi level moves depending on the doping level. In very heavily doped material, the Fermi level can move close to or even into the conduction or valence bands. Therefore, the absorption threshold could be different between the doped and undoped material. At photon energies near the bandgap, absorption could be stronger for intrinsic material compared to heavily doped material. This creates an index change due to the Kramers-Kronig relationship between doped and undoped materials. This index change can be expressed as [20,26]

$$\Delta n(\mathcal{E}) = H(\mathcal{E})N.$$

The coefficient H strongly peaks around photon energy $\mathcal{E}$ corresponding to the bandgap of the material [20]. Below the bandgap, its value is about 5×10^{-21} cm^{-3} for GaAs. This relationship holds as long as the semiconductor is not very heavily doped and the doping level is less than about 5×10^{17} cm^{-3}. For heavier doping, band tails that develop at the conduction band edge and bandgap shrinkage start to dominate the absorption. This increase in the absorption starts to give an index change that opposes the band filling effect [27]. For p-type material the variation of the Fermi level with doping concentration is much less due to a heavier hole effective mass; hence, the band-filling effect is much weaker in p type material.

4.3 PHASE MODULATORS

There are many different ways of making optical waveguides in III-V compound semiconductors. It is possible to adjust the index of refraction of these materials by controlling the composition of their alloys. For example, increasing the Al composition x in an Al$_x$Ga$_{1-x}$As compound semiconductor decreases its index of refraction. Furthermore, Al$_x$Ga$_{1-x}$As is lattice matched to GaAs for all x values. By growing such layers epitaxially using techniques such as Molecular Beam Epitaxy (MBE) or Metal Organic Chemical Vapor Deposition (MOCVD), it is possible to sandwich a higher-index material between two lower-index materials. This forms a slab waveguide and provides waveguiding in the vertical direction.

The most common approach to provide lateral waveguiding is to etch a rib and form what is known as a rib waveguide. The effective index under the rib is higher than the effective index outside the rib in the etched regions. This provides a lateral index step, and a two dimensional waveguide is formed. There are many other ways to provide waveguiding which include proton implantation, buried heterostructures, pn junctions, and disordering. However, rib waveguides are almost the universal choice since vertical and lateral index profiles and dimensions can be independently and precisely controlled. Typical rib widths are in the 2- to 4-μm range and rib heights are typically less than 1 μm. To create modulation, an electric field should be applied to the material. This is typically done using the depletion layer of reverse biased junctions. The most commonly used arrangements to apply electric fields to the material are shown in Figure 4.6.

Figure 4.6(a) uses a *pin* junction on a conducting substrate. Having a conducting substrate makes *n* ohmic contact formation and placement very easy. Usually the substrate is thinned and the back side is metalized. The *i* region is typically the core of the waveguide. If this junction is reverse biased, a large and uniform electric field is generated in the *i* region or over the core of the waveguide. However, this arrangement is not very useful for high-speed applications because the conductivity of even a heavily doped *n*-type semiconductor is not high enough to keep ohmic losses minimal at microwave and millimeter wave frequencies. This problem is eliminated using a semi-insulating (SI) substrate as shown in Figure 4.6(b). In this case an n^+ layer on a semi-insulating substrate is used to form low-resistance ohmic contacts to the *n* side of the junction. Furthermore, *n* ohmic contacts are on the front surface of the wafer, which makes the fabrication of coplanar type high-speed electrodes possible. Such heavily doped layers have large optical-loss coefficients, so the upper and lower claddings of the waveguide usually have about an order of magnitude less doping to reduce the optical loss. The thickness and composition of these layers should be chosen such that optical field should not overlap at all with the n^+ and p^+ layers and *n* and *p* metals.

For high-speed and overlap considerations, the width of the *p* metal is usually wider than the width of the rib. The wider *p* metal is usually supported by a low dielectric constant polymer such as Benzocyclobutane (BCB). A variation of this arrangement is shown in Figure 4.6(c). In this case the waveguide is etched through the core. Lateral confinement of the optical mode is very good, which in turn improves the overlap factor. For single mode operation, the width of this waveguide should be very small. Reducing the width also helps to reduce the device capacitance; however, such waveguides are difficult to make single mode since the required width is submicron. Furthermore, they suffer from excess scattering loss from the etched sidewalls. Optical propagation loss of these waveguides is around 20 dB/cm [28]. A significant part of this loss is due to doping in the material, especially *p* doping. Obviously eliminating *p* doping would reduce the optical propagation loss. Furthermore, *p* material has lower conductivity and higher microwave loss. Therefore eliminating *p* doping also helps to improve the high-speed performance.

Figure 4.6(d) shows a waveguide that supports a *nin* junction [29]. The top *p* and p^+ layers are replaced by a *n* layer and an SI InP layer. The SI InP layer is used to block the current flow between two *n* InP layers. In the InP material system it is possible to grow SI InP layers using iron doping [30]. But it is hard to duplicate the same design in the AlGaAs material system. There the approach used is to replace

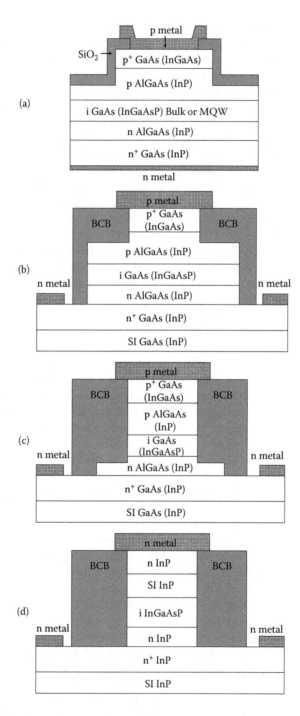

FIGURE 4.6 The most commonly used arrangements to apply electric fields to create modulation in a compound semiconductor rib waveguide.

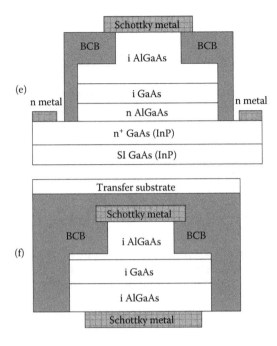

FIGURE 4.6 (Continued)

the *p* contact with a Schottky contact as shown in Figure 4.6(e). In this material system Fermi-level pinning at the surface provides a Schottky barrier height on the order of 0.7 V. Therefore, good Schottky diodes are fabricated easily. The undesirable effects of doping can be totally eliminated using undoped or unintentionally doped layers. In this case the required electric field can be applied using Schottky electrodes on either side of the epilayer as shown in Figure 4.6(f). In this case substrate should be removed to make the fabrication of electrodes on either side of the epilayer [31]. In all these cases [001]-oriented substrates are used and the applied electric field is predominantly in the [001] direction.

4.4 DRIVE VOLTAGE OF A MACH-ZEHNDER MODULATOR

In all cases, the electro-optic properties of the *i* region or the core of the waveguide should be optimized to maximize electro-optic efficiency, for example, by using a bulk or multi quantum well material of appropriate composition. When both linear and quadratic electro-optic coefficients are present and the applied electric field is in the [001] direction, the index change in the material can be expressed using Equation 4.20, Equation 4.21, Equation 4.24 and Equation 4.25 as

$$\Delta n_{TE} = \frac{1}{2} n^3 (R(\lambda) E_z^2 \pm r_{41} E_z) \qquad (4.31)$$

and

$$\Delta n_{TM} = \frac{1}{2}n^3 R(\lambda)E_z^2. \tag{4.32}$$

It should be noted that, depending on the orientation of the optical waveguide, the quadratic electro-optic effect may add or subtract from the linear electro-optic effect. The presence of the applied electric field modifies the index of the material according to Equation 4.31 and Equation 4.32. This in turn modifies the propagation constant of the optical mode. Since the effect is very small this perturbation can be found using a perturbation analysis. The result is [32]

$$\Delta\beta = \frac{2\pi}{\lambda}\Delta n_{eff} \quad \text{with} \quad \Delta n_{eff} = \iint \Delta n_{TE(TM)} \left|\Upsilon\right|^2 dS,$$

where Υ is the normalized electric field of the optical mode. The integration is carried out over the entire optical mode. Substituting the value of $\Delta n_{TE(TM)}$ derived earlier and considering only the linear electro-optic effect we obtain

$$\Delta n_{eff} = \frac{1}{2}n^3 r_{41} \iint E_z \left|\Upsilon\right|^2 dS. \tag{4.33}$$

Inside the waveguide the index of the material is not uniform. However, the index variation within the waveguide is typically small and one can replace n with the mode effective index n_{eff}. If this is not a good approximation, the required integration should be evaluated as the sum of integrations over regions of different refractive index. Multiplying and dividing this equation by $\frac{V}{g}$, where V is the applied voltage and g is the thickness of the semiconductor layer over which a high electric field exists (either the electrode gap or the thickness of the i region), we obtain

$$\Delta n_{eff} = \frac{1}{2}r_{41}n_{eff}^3 \frac{V}{g}\left[\frac{g}{V}\iint E_z \left|\Upsilon\right|^2 dS\right] = \frac{1}{2}r_{41}n_{eff}^3 \frac{V}{g}\Gamma. \tag{4.34}$$

Γ is known as the overlap integral and is expressed as

$$\Gamma = \iint \frac{E_z}{\frac{V}{g}}\left|\Upsilon\right|^2 dS. \tag{4.35}$$

Γ is proportional to the overlap of the magnitude squared normalized optical mode electric field and the normalized z component of the applied modulating field.

In a Mach-Zehnder modulator the net phase shift between the arms is given as

$$\Delta\phi = \Delta\beta_1 l - \Delta\beta_2 l = \frac{2\pi}{\lambda}l(\Delta n_{eff1} - \Delta n_{eff2}) = \frac{2\pi}{\lambda}l\left(\frac{1}{2}r_{41}n_{eff}^3 \frac{1}{g}\right)(V_1\Gamma_1 - V_2\Gamma_2) \tag{4.36}$$

where l is the length of the electrode. $\Delta\phi$ is maximized if $V_1 = -V_2$ and $\Gamma_1 = \Gamma_2 = \Gamma$. This drive condition is known as true push-pull operation. A Mach-Zehnder modulator is turned on and off when $\Delta\phi$ changes from 0 to π. The voltage required to create a π phase shift between the arms of the Mach-Zehnder modulator is known as V_π. Therefore, under true push-pull operation the voltage required to turn the modulator off is given as

$$V_\pi = \frac{\lambda g}{2lr_{41}n_{eff}^3\Gamma}.$$ (4.37)

If a strong quadratic electro-optic effect is also present Δn_{eff} can be expressed as

$$\Delta n_{eff1} = \frac{1}{2}n_{eff}^3\left[r_{41}\frac{V}{g}\Gamma_L + R\left(\frac{V}{g}\right)^2\Gamma_Q\right],$$ (4.38)

with

$$\Gamma_L = \Gamma = \iint\frac{\frac{E_z}{V}}{\frac{V}{g}}|\Upsilon|^2\,dS$$

and

$$\Gamma_Q = \iint\frac{E_z^2}{\left(\frac{V}{g}\right)^2}|\Upsilon|^2\,dS.$$

It should be noted that the definition of the overlap integrals for the linear and quadratic electro-optic effects are different. The net phase shift between the arms of a Mach-Zehnder modulator can be expressed as

$$\Delta\phi = \frac{2\pi}{\lambda}l\left(\frac{1}{2}n_{eff}^3\right)\left[\frac{r_{41}}{g}(V_1\Gamma_{L1} - V_2\Gamma_{L2}) + \frac{R}{g^2}(V_1^2\Gamma_{Q1} - V_2^2\Gamma_{Q2})\right].$$ (4.39)

$\Delta\phi$ is maximized if $V_1 = 0$ and $V_2 = V$ or $V_2 = 0$ and $V_1 = V$. In other words, using an AC signal, only one arm should be driven to get the full advantage of the quadratic effect. If $\Gamma_{L1} = \Gamma_{L2} = \Gamma_L$ and $\Gamma_{Q1} = \Gamma_{Q2} = \Gamma_Q$, then V_π is found as the positive root of the quadratic equation

$$V^2 + \frac{gr_{41}\Gamma_L}{R\Gamma_Q}V - \frac{\lambda g^2}{lRn_{eff}^3\Gamma_Q} = 0.$$ (4.40)

If V is not just an AC signal and involves a DC bias, the situation is more complicated. It is usually possible to apply a bias to both arms of the interferometer such that $V_1 = V_B + V$ and $V_2 = V_B - V$. If $|V| = V_B$ it is seen that the voltage swing across one arm will change between 0 and $2V_B$. Hence, modulating voltage will effectively be doubled. Furthermore if $\Gamma_{L1} = \Gamma_{L2} = \Gamma_L$ and $\Gamma_{Q1} = \Gamma_{Q2} = \Gamma_Q$,

$$\Delta\phi = \frac{2\pi}{\lambda} l \left(\frac{1}{2} n_{eff}^3 \right) \left[\frac{r_{41}}{g} \Gamma_L (2V) + \frac{R}{g^2} \Gamma_Q (4VV_B) \right] \qquad (4.41)$$

In this case, attention should be paid not to forward bias or break down any of the semiconductor junctions involved at any time. So $-V_{BD} < V_1 < 0$ and $-V_{BD} < V_2 < 0$, where V_{BD} is the breakdown voltage of the junction. Using Equation 4.41 we can solve for the drive voltage that will generate a π phase shift between the arms of the interferometer as

$$V_\pi = \frac{\lambda}{2 l n_{eff}^3} \frac{1}{\left[\frac{r_{41}\Gamma_L}{g} + \frac{2 R \Gamma_Q V_B}{g^2} \right]}, \qquad (4.42)$$

subject to the condition that $V_{B\,max} = V_{BD} - V_\pi$. As a matter of fact, another quadratic equation can be found to obtain the minimum V_π using this condition. It is also observed that strong wavelength dependence of R can be compensated using an appropriate bias voltage so that RV_B becomes wavelength independent. This application is discussed later.

4.5 TRAVELING WAVE ELECTRO-OPTIC MODULATOR DESIGN GUIDELINES

There is a long list of requirements for the successful utilization of optical modulators in system applications. This list definitely includes wide bandwidth, low drive voltage, low insertion loss, low wavelength and temperature sensitivity, and low cost. Some of these requirements, such as the drive voltage and bandwidth, are coupled, and usually compromises are necessary. A well-established approach to very wide electrical bandwidth modulators with low drive voltage is the so-called traveling wave design [33]. In such a design, the electrode is designed as a transmission line. Therefore, electrode capacitance is distributed and does not create an RC limit on the modulator speed. The small signal modulation response of a traveling wave modulator whose electrode is terminated by its characteristic impedance is given as [34]

$$M(f) = e^{-\left(\frac{\alpha l}{2}\right)} \left[\frac{\sinh^2\left(\frac{\alpha l}{2}\right) + \sin^2\left(\frac{\xi l}{2}\right)}{\left(\frac{\alpha l}{2}\right)^2 + \left(\frac{\xi l}{2}\right)^2} \right]^{\frac{1}{2}}, \qquad (4.43)$$

where $\xi = (n_\mu - n_o)\frac{2\pi f}{c}$, α and l are the loss coefficient and the length of the electrode, respectively. f is the electrical frequency, and c is the speed of light in vacuum. n_μ and n_o are the microwave and optical indices and are related to the microwave and optical velocities through well-known expressions. To maximize the bandwidth, microwave and optical velocities should be matched. These velocities are defined as the optical group and microwave phase velocity [35,36]. Based on Equation 4.43, if there is no velocity mismatch the 3-dB bandwidth will be at the frequency where the total electrode loss becomes 6.34 dB. Therefore, a low loss and velocity matched electrode is essential for the realization of a very wide bandwidth traveling wave modulator.

Based on this discussion and the drive voltage and bandwidth expressions given earlier, the following requirements should be satisfied to take full advantage of the traveling wave idea:

1. Propagation loss of the optical guide should be low so that a long modulator can be realized. This helps to significantly reduce the drive voltage as seen in Equation 4.37.
2. Electrode phase velocity should be matched to optical group velocity. The electrodes should not have dispersion; in other words, electrode group and phase velocities should be the same. This eliminates the velocity mismatch and the electrode can be made very long. In III-V semiconductors, velocity matching requires slowing the microwave velocity. The optical group velocity, v_g, which is the target electrode phase velocity, v_{ph}, is known for a given material system. For example, GaAs/AlGaAs material systems at 1.55 μm have a group index around 3.5 which corresponds to

$$v_g = \frac{30}{3.5} = 8.57 \text{ cm/ns} \tag{4.44}$$

Furthermore, if the desired electrode characteristic impedance is Z_0, the capacitance per unit length, C, and inductance per unit length, L, of the electrode can be expressed using the well known transmission line equations as

$$v_{ph} = \sqrt{\frac{1}{LC}}$$

and

$$Z_0 = \sqrt{\frac{L}{C}}$$

as

$$C = \frac{1}{Z_0 v_{ph}}$$

and

$$L = \frac{Z_0}{v_{ph}}.$$ (4.45)

As an example, assume a 50 Ω electrode with 8.57 cm/ns phase velocity. Then required C and L values are

$$C = 2.3 \text{ pF/cm} \quad \text{and} \quad L = 5.8 \text{ nH/cm}.$$ (4.46)

3. The electrode microwave loss coefficient should be low, so that a long modulator can be realized. As described before, total electrode loss should be lower than 6.34 dB at the desired 3-dB bandwidth point.
4. For efficient modulation, a good overlap between the optical mode and the vertical component of the microwave electric field is needed. In other words, Γ in Equation 4.37 should be as close to 1 as possible.
5. Push-pull drive should be possible to reduce the drive voltage a factor of two.
6. Electrode gap g should be as small as possible to generate as large an electric field as possible for a given voltage. However, this should be achieved without making electrode capacitance and loss (as well as optical loss) excessively large.
7. The electro-optic coefficient r_{41} should be increased as much as possible without increasing optical loss.

Specific high-speed modulator examples attempting to satisfy all or some of these requirements are given in the next section.

4.6 SPECIFIC HIGH-SPEED MODULATOR EXAMPLES

Soole et al. [37] report a phase modulator using the structure shown in Figure 4.6(a). The layer structure is p^+-InP(0.5 μm)/p-InP(0.47 μm, 3×10^{17} cm^{-3})/n-InP(0.03 μm, 1×10^{17} cm^{-3})/i-MQW/n-InP(1.5 μm, 1×10^{18} cm^{-3}) on an n^+ InP substrate. The MQW region was 0.5 μm thick. It contained 25 12-nm wide InGaAsP QWs of photoluminescence peak $\lambda = 1.37$ μm separated by 24 8-nm barriers of InP. The thin n-InP layer was used to reduce the zero-bias field across the MQW region, which allows a larger bias to be applied before breakdown occurs. Waveguides are formed by etching ~4-μm wide ridges in the InP cladding along the [1 $\overline{1}$0] direction. p ohmic contact was made by evaporating Au-Be/Au on p^+-InP ridges. n ohmic contact was formed using Au/Sn/Au pads deposited on n^+ InP through a window etched from the top surface. A thin SiO$_x$ layer was used to passivate the waveguide walls and to insulate the p contact pad from the semiconductor. During the measurements, phase modulation was converted into amplitude modulation using the Fabry-Perot cavity formed between the cleaved facets of the waveguide. This technique is commonly used to characterize phase modulators. Transmission through the Fabry-Perot cavity changes as the material index changes. Measuring the fringe visibility and the periodicity, both the index and the propagation loss of the material can be determined. The phase shift for TM polarization is given as

$$\Delta\phi_{TM}(1.55\ \mu m) = \left((2.9° \pm 0.1°)\frac{1}{V^2 - mm}\right)V^2,$$

$$\Delta\phi_{TM}(1.5\ \mu m) = \left((3.6° \pm 0.1°)\frac{1}{V^2 - mm}\right)V^2\ \text{degrees}.$$

As described earlier for this electric field orientation, there is only a quadratic electro-optic contribution for TM polarization. For TE polarization there is a linear electro-optic contribution and phase changes are given as,

$$\Delta\phi_{TE}(1.55\mu m) = \left((2.9 \pm 0.1)\frac{1}{V^2 - mm}\right)V^2 + \left((6 - 10°)\frac{1}{V - mm}\right)V$$

and

$$\Delta\phi_{TE}(1.5\ \mu m) = \left((4.1 \pm 0.1)\frac{1}{V^2 - mm}\right)V^2 + \left((6 - 10°)\frac{1}{V - mm}\right)V$$

The increased TE efficiency at 1.55 μm is due to the closer proximity of the exciton absorption edge for TE polarization to the operating wavelength. Another crucial issue for this type of modulator is the excess loss due to the change in the absorption with applied bias. This loss is proportional to the detuning between the absorption edge (~1.37 μm) and the operating wavelength. In this case 0-V propagation loss is reported to be 5.2 dB/cm and 8.6 dB/cm at 1.55 μm and 1.5 μm, respectively. Excess loss was negligible for wavelengths longer than 1.54 μm and increased linearly to 2.7 dB/cm at 1.5 μm at 5 V. The bandwidth of this modulator was predominantly limited by device and pad capacitance and was 2.7 GHz.

Another low-voltage high-speed phase modulator of the type shown in Figure 4.6(c) has recently been reported [28]. The cross section of the device is schematically shown in Figure 4.7. The device consists of 1.7-μm n-doped InP followed by a 300-nm MQW layer, a 1.5-μm p-doped InP layer and a 0.15-μm p + InGaAs cap layer.

The MQW layer, which acts as the core of the waveguide, was unintentionally doped. It had 11 QWs and had a photoluminescence peak around 1450 nm. The QWs were 80-Å wide InGaAsP with a strain of 0.75%. The barriers were InGaAsP about 170 Å thick with a strain of −0.25%. The bandgap of the well and barrier material corresponded to 1525 nm and 1300 nm, respectively. The opposite type of strain in the well and barrier material is used for strain compensation, which

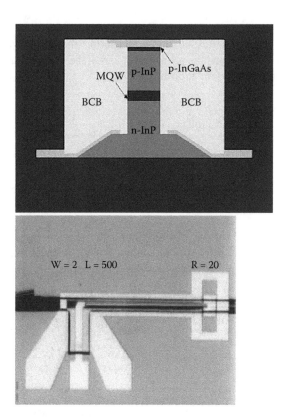

FIGURE 4.7 Cross-sectional schematic and top view of a low-voltage high-speed electro-optic phase modulator [28].

enables relatively thick stacks of strained material. The width of the waveguide shown in Figure 4.7 was approximately 2 µm and it was etched about 2.5 µm deep. Because of strong lateral confinement it is multimode. After etching and ohmic contact formation it is covered by a thick, wide BCB layer. This layer supports the center conductor of the coplanar waveguide electrode. The n contacts on either side of the waveguide are connected to the ground planes of the coplanar waveguide electrode, as seen in Figure 4.7. The electrode formed this way had a characteristic impedance of 20 Ω, as seen in Figure 4.8. Such low impedance is the result of very high capacitance of the device. At the input side, the ground planes of the coplanar waveguide were tapered out and a transition to a 20-Ω microstrip line with BCB dielectric was made. At the end of the device the coplanar waveguide electrode was terminated with a 20-Ω on-chip resistor. This serves as an on-chip matched termination.

This device is characterized using the Fabry-Perot technique described earlier. For a 500-µm long device, V_π for TE polarization was 1.8 V and 1.9 V at 1550 nm and 1560 nm, respectively. V_π for TM polarization was around 3 V. For this device the dominant physical effects that contribute to modulation are linear and

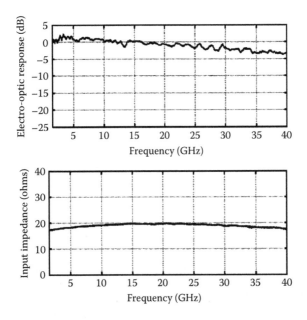

FIGURE 4.8 Small signal modulation response and the input impedance of a 500-μm long 2-μm wide phase modulator reported in [28].

quadratic-electro-optic effects. The absence of linear electro-optic effect for TM polarization results in a higher V_π. The strong wavelength dependence of TE V_π is the due to strong spectral dependence of the quadratic electro-optic effect. The propagation loss of the waveguide is reported to be 20 dB/cm. The insertion loss of the modulator, which includes coupling losses, was about 15 dB. The bandwidth of this device is reported to be 35 GHz, as seen in Figure 4.8. Due to short device length, bandwidth is probably limited by the electrode microwave loss. One of the factors that contribute to electrode loss is the p-doped layers. In compound semiconductors, p-doped material has very low conductivity. Furthermore, free carrier absorption in p layers is rather high due to intraband absorption; therefore, the p layer is also a major contributor to the propagation loss. Eliminating the p layer could improve device performance, but techniques should be devised to allow the application of high fields to the material. One of the approaches is to use an n-i-n layer with a semi-insulating current blocking layer as illustrated in Figure 4.6(d).

Such a device was recently reported [29,38]. Figure 4.9 shows the cross-sectional schematic of the waveguide used in this modulator. The core of the waveguide was an MQW layer sandwiched between 500-Å thick InGaAsP layers with bandgap corresponding to 1.3 μm. The undoped MQW region had 13 periods of 100-Å thick $In_{0.53}Ga_{0.39}Al_{0.08}As$ wells and 50-Å thick $In_{0.52}Al_{0.48}As$ barriers. The photoluminescence peak of this layer was at 1.37 μm. All the layers including the 1-μm thick current-blocking SI Fe-doped layer were grown by MOCVD. The breakdown voltage was 15 V and leakage current was less than 200 nA up to 14 V of bias. The waveguides are formed by dry etching 2-μm-wide mesas 3.5 μm deep.

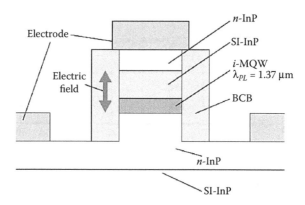

FIGURE 4.9 Cross-sectional schematic of the *nin* waveguide used in [29].

Using this waveguide, Mach-Zehnder modulators shown in Figure 4.10 were fabricated. The input and output couplers were 2×2 MMI couplers and the arms were 3 mm long. The gold electrode used was 3-μm thick and signal-to-ground spacing was 10 μm. DC transfer function of the modulator is shown in Figure 4.11. The voltage is applied only to one arm of the modulator since for the arrangement shown in

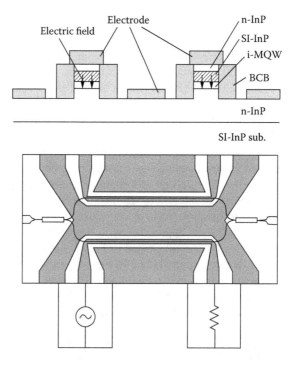

FIGURE 4.10 Cross-sectional and top schematics of the Mach-Zehnder modulator using the waveguide shown in Figure 4.9 [38].

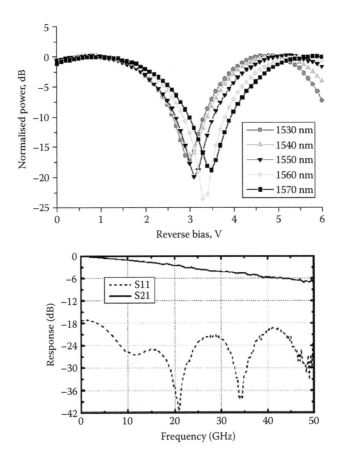

FIGURE 4.11 Optical transfer function of the modulator shown in Figure 4.10 at different wavelengths [29] and electrical s-parameters of the modulator electrode [38].

Figure 4.10, true push-pull operation is not possible. The extinction ratio and V_π were 20 dB and 2.2 V at 1550 nm. V_π shows some wavelength dependence as expected and changes about 1.2 V between 1520 nm and 1620 nm. Between 1530 nm and 1570 nm the extinction ratio is better than 13 dB for a 2-V voltage swing. The change in the absorption was also negligible due to large detuning between the MQW absorption peak and the operating wavelength. This is observed by comparing optical output at 1 V at 2 V_π. The s-parameters of the modulator electrode are shown in Figure 4.11; s_{11} is less than −17 dB, which indicates that the electrode impedance is very close to 50 Ω. In this case the 3-dB electrical modulation bandwidth occurs when s_{21} equals −6.34 dB; hence, it occurs at 42 GHz. This modulator was used in a transmission experiment, and clear eye opening is observed at 40 Gbits/sec with a drive voltage of 2.3 V supplied to one arm.

There are two difficulties with this approach. One is the difficulty of obtaining true push-pull drive. The other is the variation of the drive voltage with operating wavelength. Not having a push-pull drive increases the chirp of the modulator and

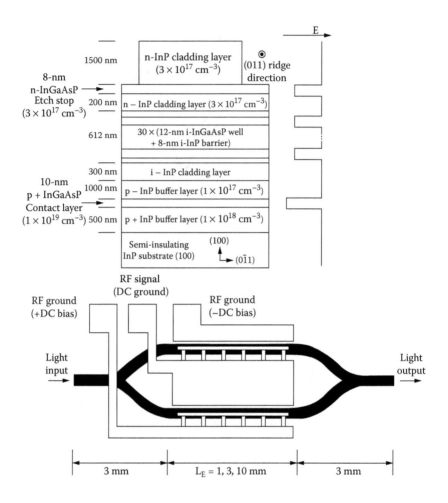

FIGURE 4.12 The material design and top schematic of the modulator reported in [39].

doubles the drive voltage. This difficulty can be circumvented by employing a DC bias common to both arms. This approach is used in [39].

The material design and top schematic of the modulator reported in [39] is shown in Figure 4.12. The core of the waveguide contains 30 pairs of 120-A InGaAsP QWs and 80-A InP barriers. The photoluminescence wavelength of the QWs was 1420 nm. In this design the p layer was at the bottom to allow larger area p contacts which in turn helps to reduce the resistance due to p contacts. The optical waveguide is obtained by etching the top InP layer to the InGaAsP etch stop layer. The width of the waveguides varied from 2 to 4 μm. Then a deep ridge is etched to the top of the p + InGaAs contact layer. The ridge formed this way was 12 μm wide. Finally isolation between modulators was obtained by etching to the SI InP substrate. In the biasing scheme shown in Figure 4.12 it is possible to maintain a reverse bias across both arms. If the bias voltage magnitude is V_b and the RF signal swings between $+V_b$ and $-V_b$ the net voltage across each arm swings between 0 and $2V_b$. Therefore, the voltage that appears across an arm of the interferometer is twice the RF voltage amplitude, and efficiency of the modulation is increased by a factor of two. This particular modulator had a V_π of 0.9 V for a 1-cm long electrode. Total length of the device was 16 mm. The on chip propagation loss was given as 7 dB/cm. The bandwidth was 1.2 GHz under lumped drive conditions.

DC bias can also be used to eliminate the wavelength dependence of the drive voltage experienced in modulators using the quadratic electro-optic effect. In the quadratic electro-optic effect, the index change is more effective at shorter wavelengths closer to the exciton peak. At the same time, the index change is proportional to the square of the applied field; hence, large index changes are obtained at higher fields. These two properties can be used together to eliminate the change in the required drive voltage as the operating wavelength changes. If one uses a bias field plus an AC field, the net field change can be adjusted by changing the bias field while keeping the AC field fixed. This can be seen clearly by remembering that when both linear and quadratic electro-optic effects are present,

$$\Delta n_{TE} = \frac{1}{2} n^3 (R(\lambda) E_z^2 + r_{41} E_z).$$

If the voltage is applied to both arms of an interferometer such that $V_1 = V_B + V_{AC}$ and $V_2 = V_B - V_{AC}$, the electric fields across both arms can be written as $E_1 = E_B + E_{AC}$ and $E_2 = E_B - E_{AC}$. Then when $E_{AC} = 0$ there is no differential phase shift between the arms. When $E_{AC} = E_{AC\,max}$ the maximum differential phase shift $\Delta\phi_{max}$ is obtained. Using the index change described above we can express $\Delta\phi_{max}$ as

$$\Delta\phi_{max} = \frac{2\pi l}{\lambda} \frac{1}{2} n^3 \{ R(\lambda)[(E_B + E_{AC\,max})^2 - (E_B - E_{AC\,max})^2]$$

$$+ r_{41}[(E_B + E_{AC\,max}) - (E_B - E_{AC\,max})]\}$$

$$= \frac{\pi l}{\lambda} n^3 [R(\lambda)(4 E_B E_{AC\,max}) + 2 r_{41} E_{AC\,max}]$$

It is clearly observed that changes in $R(\lambda)$ can be compensated by adjusting E_B. This eliminates strong wavelength dependence due to the $R(\lambda)$ term and $\frac{1}{\lambda}$ dependence is negligible over a narrow wavelength range where the quadratic electro-optic effect is strong. This approach is used in [40] and [41]. Top and cross-sectional schematics of the modulator operated using this principle are shown in Figure 4.13.

Figure 4.14 shows the variation of V_π as a function of bias voltage at different wavelengths. At a fixed bias voltage, as wavelength increases the quadratic electro-optic effect weakens. As a result, V_π increases. At a fixed wavelength, increasing bias voltage increases the applied field and reduces V_π through the quadratic electro-optic effect. This observation shows that the reduction in $R(\lambda)$ for increasing wavelength can be compensated by increasing bias voltage. This is equivalent to drawing

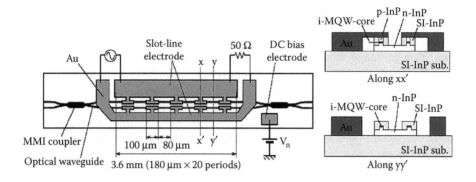

FIGURE 4.13 Top and cross-sectional schematics of the Mach-Zehnder modulator described in [40].

a horizontal line in Figure 4.14 at the desired V_π and choosing the appropriate bias voltage at the desired wavelength.

The measured transfer function obtained this way at different wavelengths is also shown in Figure 4.14. In this case the bias voltage had to be adjusted from 3 V to 9.8 V as the wavelength changed from 1.51 μm to 1.57 μm. Although the bias voltage range was rather wide, it was possible to keep the required AC drive voltage to turn the modulator on at 1.9 V. The dynamic chirp of this device was also measured under 10 Gb/s drive conditions. The chirp is reported to be within ± 0.8 GHz and the chirp parameter was + 0.1 [41]. This device is used in 10 Gb/s transmission experiments. Another important feature of this device is the design of the microwave electrodes. It uses segmented electrodes to reduce the microwave loss and the capacitive loading of the electrode. Such designs are commonly used to improve the bandwidth of the modulator and are discussed in the next section.

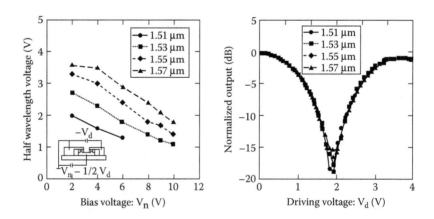

FIGURE 4.14 Measured V_π as a function of bias voltage at different wavelengths and transfer function of the modulator reported in [40].

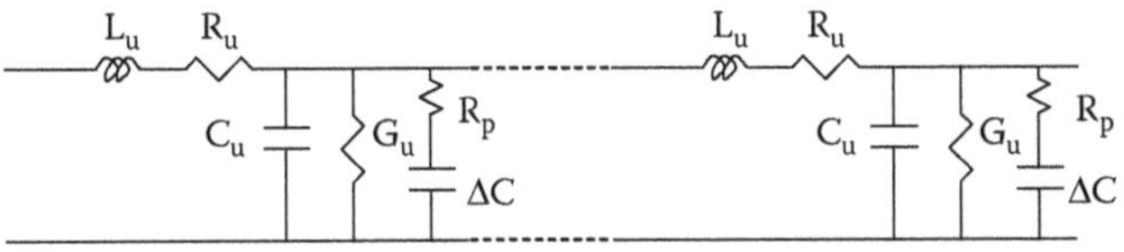

FIGURE 4.15 The electrical equivalent circuit of the capacitively loaded traveling-wave electrode [43].

In such designs the main electrode is an unperturbed microwave transmission line. In the example shown in Figure 4.13, the main electrode is a coplanar strip line. This electrode is periodically loaded by narrow and small capacitive elements. The periodical loading of the unloaded line with these tiny capacitors form a slow wave transmission line [42]. The equivalent circuit of the resulting electrode is shown in Figure 4.15. L_u, C_u, R_u, and G_u are the inductance, capacitance, resistance, and conductance per unit length of the unloaded transmission line. ΔC is the capacitance per unit length due to capacitive loading on one arm of the modulator, and R_p is due to the resistance of the stems connecting capacitive loading elements to the main line.

The capacitive loading elements are electrically isolated from one another. In the device reported in [40] and shown in Figure 4.13, electrical isolation is achieved by growing semi-insulating InP layers on top of the core in sections that are not connected to the main electrode. Therefore, transmission line currents cannot flow along the arms of the modulator due to segmentation. Only charge is pumped in and out of the capacitive elements in the form of displacement current. Axial transmission line currents actually flow along the unloaded transmission line sections where current magnitude is less due to lower electric field. As a result, ohmic losses remain essentially unchanged. There is only a slight increase due to the resistance of the stems and the displacement current through the capacitive loading elements. Hence, such loaded line designs provide very high electric fields without increasing the electrode microwave loss significantly.

As discussed before, the presence of *p* layers can increase optical loss and electrode resistance. This difficulty can be eliminated replacing the *p-i-n* junction with a Schottky-*i-n* junction. This idea results in the waveguide shown in Figure 4.6 (e). Figure 4.16 shows the schematic of a traveling-wave Mach-Zehnder modulator using such a waveguide design [44]. The optical structure is a Mach-Zehnder interferometer utilizing multimode interference sections at the input and output for power splitting and combining. A GaAs/AlGaAs epitaxial layer is grown on a semi-insulating (SI) GaAs substrate. Underneath the GaAs core there is a buried n⁺ layer which acts as a ground plane. The electrode is capacitively loaded traveling-wave type and the capacitive elements used are the reverse-biased capacitance of a Schottky-i-n junction as shown in Figure 4.16. The advantage of this approach is to utilize the large vertical electric field existing in the reverse-biased Schottky-i-n junction. This field overlaps very well with the optical mode, enabling low drive voltages. Furthermore, these capacitive elements do not carry any of the axial currents in the transmission line. Therefore, the resistance per unit length and propagation loss of

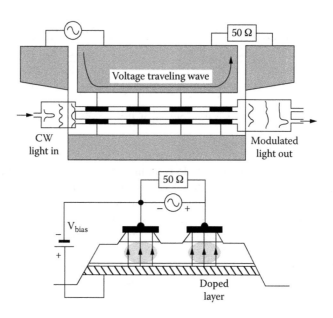

FIGURE 4.16 Schematic of traveling wave Mach-Zehnder modulator using capacitively loaded coplanar strip line [44].

the loaded line is very close to that of the unloaded line. In this design the required loading capacitance can be expressed as [44]

$$C_L = \frac{n_o^2 - n_\mu^2}{cZ_0 n_o},$$
(4.47)

where n_o and n_μ are the optical and microwave indices, c is the speed of light in a vacuum, and Z_0 is the characteristic impedance of the loaded line. In GaAs for a coplanar electrode, a loaded line impedance of 50 Ω, and using the previously quoted values, one obtains $C_L \cong 0.9$ pF/cm. This value is about 4 to 5 times smaller than the capacitance of a typical reverse-biased Schottky-i-n junction. Therefore, to obtain impedance and velocity matching the length of the loading elements needs to be reduced. This makes only a fraction of the total electrode length electro-optically active, increasing the drive voltage of the modulator. Part of this difficulty can be circumvented using a series push-pull drive illustrated in Figure 4.17. In this case arms of the Mach-Zehnder interferometer are connected in series. Hence, half the applied voltage drops across each arm but total loading capacitance is also half of the loading capacitance of each arm. This bias scheme also requires bias and decoupling circuitry, as illustrated in Figure 4.17. It is not possible to use a full or parallel push-pull bias in this scheme because of the continuous n+ layer underneath the arms of the interferometer. Such devices with a total of 1-cm electrode length and loading segment lengths of 0.5, 0.6, and 0.7 mm were fabricated and

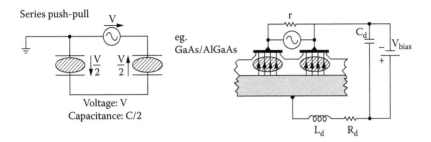

FIGURE 4.17 Series push-pull drive for the modulator shown in Figure 4.16. Phase modulator arms are modeled as capacitors [44].

characterized [44]. DC V_π values at 1.15 μm for 0.5, 0.6, and 0.7 mm segment-length electrodes were 5.7, 4.25, and 4.24 V, respectively. Corresponding bandwidths were > 26.5, 25.0, and 22.5 GHz, respectively. As the segment length gets longer, more of the electrode becomes electro-optically active and drive voltage reduces. On the other hand, capacitive loading becomes excessive and it becomes difficult to obtain impedance and velocity matching. As a result the device bandwidth decreases. Such slow-wave electrodes are inherently low-pass filters. The upper end of the passband is determined by the Bragg frequency of the slow-wave geometry and depends on the periodicity of loading elements. It is possible to push this frequency to very high values by making the period of loading elements very short. It is observed that for long-period devices, the upper frequency cutoff becomes very sharp and abrupt. On the other hand, short-period devices have a much more gradual frequency roll-off [44]. Recently such a device demonstrated an electrical 3-dB bandwidth of 50 GHz and V_π of 13 V at 1530 nm [45].

The remaining drawbacks of this type of design are the doping and the lack of true push-pull operation. Eliminating doping altogether further reduces the optical and microwave losses. Recently a novel GaAs traveling wave modulator design using unintentionally doped epitaxial layers was reported [46,47]. Such unintentionally doped GaAs/AlGaAs layers self-deplete due to Fermi-level pinning at the surface and the depletion originating at the semi-insulating substrate interface and behave very similar to low-loss dielectric materials. As a result, optical and microwave losses become very low. The required velocity slowing can be achieved using a properly designed electrode. A schematic of such a device is shown in Figure 4.18. The optical structure is a Mach-Zehnder interferometer. A (001)-oriented GaAs/AlGaAs heterostructure grown by MBE on semi-insulating GaAs provides vertical optical waveguiding. The lateral waveguiding is obtained by etching ridges down into the top AlGaAs layer. The microwave electrodes are fabricated by evaporating 200Å/200Å/1μm of Ti/Pt/Au. This forms a Schottky contact with the epitaxial layers. A voltage applied between the electrodes biases two back-to-back Schottky diodes. The conductivity of self-depleted epilayers is so low that, for frequencies larger than 1 MHz or even lower, the epilayers start to behave like slightly lossy dielectrics. Hence, the situation becomes identical to electrodes on an insulating dielectric.

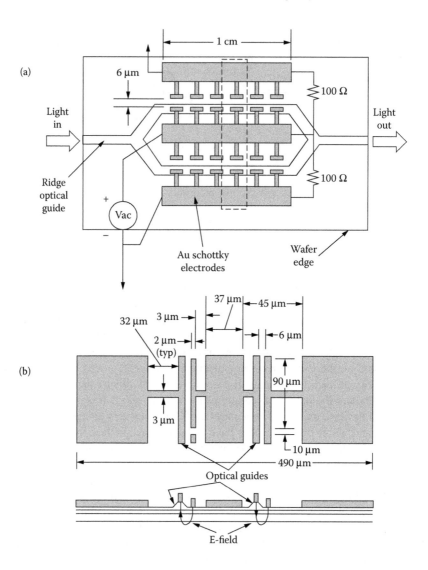

FIGURE 4.18 (a) Schematic top view of a GaAs/AlGaAs modulator using unintentionally doped layers. (b) Top view schematic of the modulator section delineated by the dashed line in (a) together with a cross-sectional schematic [46].

This makes it possible to apply mainly (001)-directed electric fields of opposite polarity on the optical guides, as shown in Figure 4.18. This generates phase shifts of opposite sign on both arms through the linear electro-optic effect, creating a net differential phase shift between the arms of the interferometer. Hence, true push-pull modulation results. The device electrode is a modified coplanar line, in which T-rails stem from either side of the center conductor and from the inner side of both ground planes [48]. These T-rails form tiny capacitors, which periodically load the line as discussed earlier.

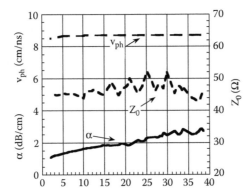

FIGURE 4.19 Measured characteristics of the modulator electrode used in the modulator reported in [46] and [48].

Figure 4.19 shows the measured characteristics of the modulator electrode used in the modulator reported in [46] and [48]. In this case the small gap between the T-rails is 6 μm. The measured phase velocity shows virtually no dispersion and is within 1% of the target value. The characteristic impedance is 46 Ω, which is very close to the target value of 50 Ω. The microwave loss is also very low. It increases with frequency and approaches a value of 3 dB/cm at 40 GHz. This value is considerably less than the 6.34-dB loss limit that determines the 3-dB electrical bandwidth when velocity and impedance matching is obtained. Therefore, a very high electrical bandwidth is expected from this device. Another advantage of this design is the symmetry of the high field region and the large segmentation factor. The 6-μm gap extends over 90% of the electrode length. The increase in the microwave loss of the unperturbed coplanar line due to the presence of the T-rails is measured to be less than 1 dB over the entire frequency range. It is possible to reduce the gap further. A T-rail gap of 3 μm results in a microwave loss of 4.3 dB/cm at 35 GHz [48]. Although loss increases, its absolute value is still very low. Such small gap values make this geometry suitable for both directional coupler and Mach-Zehnder type electro-optic modulators.

The small-signal electrical bandwidth of this device at 1.55 μm is shown in Figure 4.20. The bandwidth at 1.55 μm is in excess of 40 GHz. It is basically flat up to 20 GHz and starts to roll off gradually and becomes about 1.5 to 2 dB down at 40 GHz. Extrapolating the curve fit, the bandwidth was estimated to be between 50 to 60 GHz. Although this is a rather high bandwidth it is still less than the bandwidth expected based on the electrode data. At 1.3 μm, the 3-dB bandwidth was measured to be about 37 GHz [47,49]. This is considerably less than what was expected. The reason for this discrepancy was shown to be the matching of wrong velocities, namely matching of phase velocity rather than group velocity. This makes sense because once the optical signal starts to interact with the electrical signal, it is no longer a single-frequency waveform. It is clearly phase modulated and this phase-modulated waveform travels with the group velocity. This velocity should be the same as the group velocity of the electrical waveform so that they keep interacting

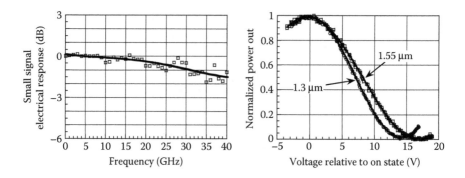

FIGURE 4.20 Measured small signal response at 1.55 μm and DC transfer function at two different wavelengths of the modulator in [46].

and the phase modulation accumulates. Therefore, group velocities should be matched. Since the electrode is a quasi-TEM transmission line, it has no dispersion, and phase and group velocities are the same. This is also obvious from the data. However, for the optical guide, phase and group velocities are different. The actual group velocity mismatch turns out to be 10% at 1.3 μm and 5% at 1.55 μm, which explains the measured bandwidths. Actually, this experimental observation is consistent with the claims made in one of the early publications on this topic [50]. This is a very important point especially for semiconductor modulators where material index dispersion is much more significant especially close to the band edge. Providing the right kind of velocity match is very important to get the widest possible bandwidth in traveling-wave modulators.

The V_π of this modulator was 14 V at 1.3 μm and 16.8 V at 1.55 μm, as shown in Figure 4.20. These values are high for a high-speed modulator. The source of this difficulty is the poor overlap of the vertical component of the applied electric field with the optical mode. As the electrode gap gets smaller, the applied field intensity increases. However, field lines start to become more tangential to the surface and the increase in its vertical component is not as great.

This difficulty can be eliminated by using substrate-removed waveguides [51] and processing both sides of the epilayer. The resulting optical waveguide becomes as shown in Figure 4.6(f). High-performance electro-optic modulators using substrate removal and hybrid integration techniques have recently been demonstrated [43,52,53,54]. Figure 4.21 shows the various schematics of one of those modulators [54]. The optical part is a Mach-Zehnder interferometer. The arms of the interferometer are made out of the semiconductor epilayer removed from its substrate. Each arm contains a single-mode optical waveguide. These arms are glued to a transfer substrate using the polymer Benzocyclobutane (BCB). In this case, the transfer substrate is semi-insulating GaAs, but other optically flat surfaces can also be used. The electrode is a modified coplanar waveguide (CPW). There are narrow Schottky electrodes at the top and bottom of the optical waveguides. These electrodes are segmented along the optical waveguide and connected to the ground planes and the center conductor of the CPW using short stems as shown in Figure 4.21.

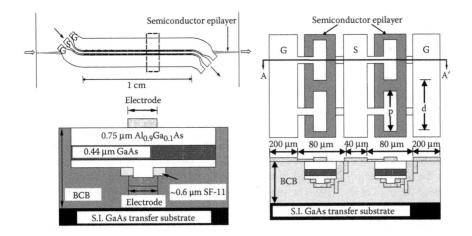

FIGURE 4.21 Top and cross-sectional schematics of a recently fabricated substrate removed modulator along with the cross-sectional profile of the optical waveguide with top and bottom T-rails [54].

They are called T-rails due to their shapes. A small capacitance is formed between the T-rails on the top and the bottom of the optical waveguide. These small capacitances are electrically connected between the electrodes of the CPW in a periodic fashion. This increases the capacitance per unit length of CPW. Its inductance per unit length remains unchanged. Its resistance per unit length increases slightly due to resistance of the short stems which carry the displacement current through small capacitors between T-rails. As described before, such periodic capacitive loading creates a slow wave electrode. The period of loading was chosen as 100 μm. This assures that the Bragg frequency is high enough to prevent low-pass filter behavior. Therefore, at frequencies up to 100 GHz, the phase velocity v_{ph} and characteristic impedance Z_0 of the resulting electrode can be expressed as

$$v_{\mu} = \frac{1}{\sqrt{L(C+2\Delta C)}}, \quad Z_0 = \sqrt{\frac{L}{C+2\Delta C}}, \quad (4.27)$$

where

$$\Delta C = K \frac{\varepsilon_{GaAs} r}{t} F \quad \text{and} \quad F = \frac{p}{d}.$$

L and C are the inductance and capacitance per unit length of the main CPW, and ΔC is the capacitance per unit length per arm between T-rails. ΔC is given as $\Delta C = K \frac{\varepsilon_{GaAs} r}{t} F$, where r is the rail width, t is the epithickness, K is a geometric factor accounting for fringing fields and dielectric nonuniformity, p is the length of a T-rail, and d is the period. Furthermore, the increase of the microwave loss of the unloaded line due to capacitive loading is very slight. Figure 4.21 shows the details of one of the optical waveguides along with the Schottky T-rail electrodes. In this design, since the arms of the interferometer can be independently biased the chirp is adjustable. True push-pull operation with zero chirp is possible. The epitaxial layer is unintentionally doped. Since it has very low background doping, it self-depletes due to Fermi-level pinning at its surfaces. Therefore the remaining semi-conductor can be treated as a slightly lossy dielectric with metal electrodes on either side. This is consistent with our previous observations [52].

In this design the electrode gap is the same as the epilayer thickness. In order to reduce the drive voltage, the epilayer thickness should be kept as low as possible. On the other hand, the optical mode should not overlap with the metal electrodes to keep the optical loss low. The vertical confinement or the index difference between the core and the cladding should be increased as much as possible. The design shown in Figure 4.21 satisfies these requirements while keeping the epilayer thickness slightly below 2 μm. Details of this study and electrode optimization using the finite element method are described in [43].

Figure 4.22(a) shows the optical transfer function of three types of modulators. The velocity-matched modulator with 50-Ω impedance has a fill factor of 35%. Such a low fill factor is needed to keep capacitive loading low enough to satisfy the velocity-matching condition while keeping the impedance 50 Ω. Figure 4.22(b) shows the electrode properties of this device. In this figure, the spikes in the imped-ance plot are due to the use of a 97-Ω calibration impedance. Whenever the small reflections from both ends of the electrode go out of phase and cancel, the electrode appears to be perfectly matched; that is, it appears to have 97-Ω impedance. Based on velocity matching and low microwave loss, this device has a bandwidth in excess of 40 GHz. This modulator has a V_π of about 10 V. If the fill factor is increased to

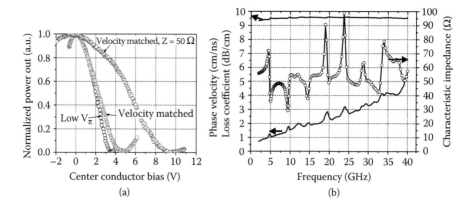

FIGURE 4.22 (a) Normalized transfer function of different types of modulators. (b) Microwave phase velocity, loss coefficient, and characteristic impedance of the velocity matched modulator with 50 Ω impedance as a function of frequency. For all modulators the electrode length is 1 cm. In (b) the spikes in the impedance plot are due to the use of a 97-Ω calibration impedance [54].

70%, V_π goes down to about 5 V, but only velocity can be matched. Impedance goes down to about 40 Ω [54]. Increasing the fill factor to 95% decreases V_π to 3.7 V. This is the lowest value reported for an undoped bulk GaAs electro-optic modulator [53]. However, in this case if velocity is matched impedance goes down to 18 Ω [43].

4.7 SUMMARY

The field of III-V compound semiconductor electro-optic modulators is an active and promising research area. It is possible to use the linear and quadratic electro-optic properties of these materials to realize very efficient and fast optical modulators. Quadratic electro-optic properties can be enhanced significantly using multiquantum wells of appropriate thickness and composition. Data and design formulas to determine these coefficients are presented. Although the electro-optic coefficients in this material system are lower than what is possible in ferroelectric materials, the high index of refraction and low dielectric constant dispersion help to fabricate very efficient modulators. It has also been shown that many different types of modulators can be fabricated using very advanced semiconductor processing techniques. Formulations are presented to obtain the drive voltage of Mach-Zehnder type intensity modulators using both linear and quadratic electro-optic effects. Traveling wave electro-optic design guidelines are also given. This is followed by specific high speed modulator examples. Using the appropriate design, it is possible to make 500-μm long phase modulators with less than 2-V drive voltage and more than 40-GHz bandwidth. Furthermore, the wavelength dependence of the quadratic electro-optic coefficients can be compensated by adjusting the DC bias. The drive voltage can be reduced to below 1 V for electrode length of 1 cm. This value is significantly lower than what can be achieved in $LiNbO_3$ even with advanced processing techniques.

Another advantage of this material system is the ability to use the loaded-line traveling-wave design. Since velocity matching in this material system requires slowing the microwave electrical signal, a regular electrode periodically loaded with small capacitances is possible. These small capacitances can be used to apply very high modulating fields which overlap very well with the optical mode. This is the basic design used in very high-speed modulators. Design guidelines for these types of modulators are given and it has been shown that bandwidths in excess of 40 GHz are readily available. Furthermore, it is possible to use substrate removal and polymer integration techniques to improve the drive voltage and bandwidth of such modulators even with bulk GaAs material. Improving this approach, it should be possible to fabricate modulators with less than 1 V drive voltages and more than 40 GHz electrical bandwidth. These developments along with the ability to integrate electronic drivers with the modulators are expected to yield low-cost, high-performance components for future photonic systems.

REFERENCES

1. J. F. Nye, Physical Properties of Crystals, Oxford, 1964.
2. M. Born and E. Wolf, Principles of Optics, Pergamon Press, 1980.
3. I. P. Kaminow, An Introduction to Electro-optic Devices, Academic Press, 1974.

4. M. Glick, D. Pavuna, and F. K. Reinhart, "Electro-optic effects and electroabsorption in a GaAs/AlGaAs multiquantum-well heterostructure near the bandgap," Electron. Lett., 23, 1235, 1987.

5. S. S. Lee, R. W. Ramaswamy, and V. S. Sundaram, "Analysis and Design of High-Speed High-Efficiency GaAs-AlGaAs Double Heterostructure Waveguide Phase Modulator," IEEE J. Quantum Electron., QE-27, 726–736, 1991.

6. N. Suzuki and K. Tada, "Electrooptic Properties and Raman Scattering in InP," Jpn. J. Appl. Phys., 23, 291–295, 1984.

7. J. Faist and F. K. Reinhart, "Phase Modulation in GaAs/AlGaAs double Heterostructures II, Experiment," J. Appl. Phys., 67, 7006–7012, 1990.

8. M. Glick, F. K. Reinhart, and D. Martin, "Linear electro-optic effect: comparison of GaAs/AlGaAs multi-quantum-well heterostructures with an AlGaAs solid solution at 1.1523 μm," J. Appl. Phys., 63, 5877– 5879, 1988.

9. C. A. Berseth, C. Wuethrich, and F. K. Reinhart, "The Electro-optic Coefficients of GaAs: Measurements at 1.32 and 1.52 μm and the Study of Their Dispersion Between 0.9 and 10 μm," J. Appl. Phys., 71, 2821–2825, 1992.

10. M. Sugie and K. Tada, "Measurement of the Linear Electro-optic Coefficients and Analysis of the Nonlinear Susceptibilities in Cubic GaAs and Hexagonal CdS," Jpn. J. Appl. Phys., 15, 421–431, 1976.

11. A. Yariv, C. A. Mead, and J. V. Parker, "GaAs as an electrooptic modulator at 10.6 microns." IEEE J. Quantum Electron., QE-2, 243–245, 1966.

12. W. D. Johnson and I. P. Kaminow, "Contributions to Optical Nonlinearity in GaAs as Determined from Raman Scattering Efficiencies," Phys. Rev., 188, 1209–1211, 1969.

13. I. P. Kaminow, "Measurements of the electrooptic effect in CdS, Zn Te, and GaAs at 10.6 microns," IEEE J. Quantum Electron., QE-4, pp. 23–26, 1968.

14. H. G. Bach, J. Krauser, H. P. Nolting, R. A. Logan, and F. K. Reinhart, "Electro-optical Light Modulation in InGaAsP/InP Double Heterostructure Diodes." Appl. Phys. Lett., 42, 692–694, 1983.

15. A. Hernandez-Cabrera, C. Tejedor, and F. Meseguer, "Linear electro-optic effects in zinc blende semiconductors," J. Appl. Phys., 58, 4666–4669, 1985.

16. S. Adachi and K. Oe, "Linear Electro-Optic Effects in Zincblende-type Semiconductors: Key Properties of InGaAsP Relevant to Device Design," J. Appl. Phys., 56, 74–80, 1984.

17. W. Franz, Z. Naturforsch., 13a, 484, 1958.

18. L. V. Keldysh, Sov. Phys. JETP, 7, 788, 1958.

19. J. Faist, F. K. Reinhart, D. Matin, and E. Tuncel "Orientation Dependence of Phase Modulation in a p-n Junction GaAs/Al$_x$Ga$_{1-x}$As Waveguide," Appl. Phys. Lett., 50, 68–70, 1987.

20. J. G. Mendoza-Alvarez, L. A. Coldren, A. Alping, R. H. Yan, T. Hausken, K. Lee, and K. Pedrotti, "Analysis of Depletion Edge Translation Lightwave Modulators," J. Lightwave Technol., 6, 793–808, 1988.

21. J. F Vinchant, J. A. Cavailles, M. Erman, P. Jarry, and M. Renaud. "InP/GaInAsP Guided-wave Phase Modulators Based on Carrier-induced Effects: Theory and Experiment," J. Lightwave Technol., 10, 63–70, 1992.

22. A. Alping and L. A. Coldren, "Electrorefraction in GaAs and InGaAsP and its Application to Phase Modulators," J. Appl. Phys., 61, 2430–2433, 1987.

23. B. R. Bennett and R. A. Soref, "Electrorefraction and Electroabsorption in InP, GaAs, GaSb, InAs and InSb," IEEE J. Quantum Electron., QE-23, 2159–2166, 1987.

24. S. Adachi and K. Oe, "Quadratic Electro-optic (Kerr) Effects in Zincblende-type Semiconductors: Key Properties of InGaAsP Relevant to Device Design," J. Appl. Phys., 56, 1499–1504, 1984.

25. J. E. Zucker, I. Bar-Koshep, B. I. Miller, U. Koren, and D. S. Chemla, "Quaternary Quantum Wells for Electro-optic Intensity and Phase Modulation at 1.3 and 1.55 μm," Appl. Phys. Lett., 54, 10–12, 1989.

26. J. G. Mendoza-Alvarez, R. H. Yan, and L. A. Coldren "Contribution of the Band Filling Effect to the Effective Refractive Index Change in DH GaAs/AlGaAs phase Modulators," J. Appl. Phys., 62, No. 11, pp. 4548–4553, December 1, 1987.

27. J. G. Mendoza-Alvarez, F. D. Nunes, and N. B. Pate, "Refractive Index Dependence on Free Carriers for GaAs," J. Appl. Phys., 51, 4365–4367, 1980.

28. L. Zhang, J. Sinsky, D. Van Thourhout, N. Sauer, L. Stulz, A. Adamiecki, and S. Chandrasekhar, "Low Voltage High Speed Travelling Wave InGaAsP/InP Phase Modulator," IEEE Photonics Technol. Lett., 16, 1831–1833, 2004.

29. K. Tsuzuki, T. Ishibashi, T. Ito, S. Oku, Y. Shibata, R. Iga, Y. Kondo, and Y. Tohmori, "40 Gbit/s n–i–n InP Mach–Zehnder modulator with a π voltage of 2.2 V," Electronics Lett., 39, 1464–1466, 2003.

30. P. J. Corvini and J. E. Bowers, "Model of trap filling and avalanche breakdown in semi-insulating Fe:InP," J. Appl. Phys., 82, 259–269, 1997.

31. S. R. Sakamoto, C. Ozturk, Y. T. Byun, J. Ko, and N. Dagli, "Low Loss Substrate-Removed (SURE) Optical Waveguides in GaAs/AlGaAs Epitaxial Layers Embedded in Organic Polymers," IEEE Photonics Technol. Lett., 10, 985–987, 1998.

32. L. A. Coldren and S. W. Corzine, Diode Lasers and Photonic Integrated Circuits, Wiley Interscience, 1995.

33. N. Dagli, "Wide Bandwidth Lasers and Modulators for RF Photonics," IEEE Trans. Microwave Theory Tech., MTT-47, 1151–1171, 1999.

34. K. Kubota, J. Noda, and O. Mikami, "Traveling Wave Optical Modulator Using a Directional Coupler LiNbO$_3$ Waveguide," IEEE J. Quantum Electron., QE-16, 754–760, 1980.

35. R. Spickermann, S. R. Sakamoto, and N. Dagli, "In Traveling Wave Modulators Which Velocity to Match?" Proceedings of IEEE/LEOS' 96, 9th Annual Meeting, Paper WM3, Boston, MA, November 18–21, 1996, pp. 97–98.

36. R. Spickermann, S. R. Sakamoto, and N. Dagli, "GaAs/AlGaAs Traveling Wave Electrooptic Modulators," Proceedings of Optoelectronic Integrated Circuits Conference, Paper 33, SPIE International Symposium on Optoelectronics' 97, Vol. 3006, pp. 272–279, San Jose, CA, February 8–14, 1997.

37. J. B. D. Soole, H. K. Tsang, I. H. White, H. P. Leblanc, R. Bhat, and M. A. Koza "High Performance QCSC Phase Modulator for the 1.5–1.55 μm Fibre Band," Electron. Lett., 26, 1421–1423, 1990.

38. K. Tsuzuki, T. Ishibashi, T. Ito, S. Oku, Y. Shibata, T. Ito, R. Iga, Y. Kondo, and Y. Tohmori, "A 40-Gb/s InGaAlAs–InAlAs MQW n-i-n Mach–Zehnder Modulator With a Drive Voltage of 2.3 V," IEEE Photonics Technol. Lett., 17, 46–48, 2005.

39. P. W. Juodawlkis, F. J. O'Donnell, R. J. Bailey, J. J. Plant, K. G. Ray, D. C. Oakley, A. Napoleone, and G. E. Betts, "Sub-Volt-V$_\pi$ InGaAsP Electrorefractive Modulators Using Symmetric, Uncoupled Quantum Wells," Proceedings of IEEE/LEOS'03, 16th Annual Meeting, Paper ThC5, Tucson, AZ, October 26–30, 2003, pp. 27–28.

40. Suguru Akiyama, Hiroaki Itoh, Tatsuya Takeuchi, Akito Kuramata, and Tsuyoshi Yamamoto, "Wide-wavelength-band (30 nm) 10-Gb/s Operation of InP-based Mach-Zehnder Modulator with Constant Driving Voltage of 2 Vpp," IEEE Photonics Technol. Lett., 17, 6, 2005.

41. S. Akiyama, H. Itoh, T. Takeuchi, A. Kuramata, and T. Yamamoto, "Low-chirp 10-Gb/s InP-based Mach-Zehnder modulator driven by 1.2 V single electrical signal," Electron. Lett., 41, 40–41, 2005.

42. R. E. Collin, Foundations for Microwave Engineering, McGraw Hill, 1966.
43. J. H. Shin, C. Ozturk, S. R. Sakamoto, Y. J. Chiu, and N. Dagli, "Novel T-Rail Electrodes for Substrate Removed Low-Voltage, High-Speed GaAs/AlGaAs Electro-optic Modulators," IEEE Trans. Microwave Theory Tech., MTT-53, 636–643, 2005.
44. R. G. Walker, "High speed III-V electrooptic waveguide modulators," IEEE J. Quantum Electron., 27, 654–667, 1991.
45. R. G. Walker, "Electro-optic Modulation at mm-wave Frequencies in GaAs/AlGaAs Guided Wave Devices," Proceedings of IEEE/LEOS'95, 8th Annual Meeting, paper IO4.2, San Francisco, CA, October 30–November 2, 1995, pp. 118–119.
46. R. Spickermann, S. R. Sakamoto, M. G. Peters and N. Dagli, "GaAs/AlGaAs Traveling Wave Electrooptic Modulator with Electrical Bandwidth greater than 40 GHz," Electron. Lett., 32, 1095–1096, 1996.
47. R. Spickermann, S. R. Sakamoto and N. Dagli, "GaAs/AlGaAs Traveling Wave Electrooptic Modulators," Proceedings of Optoelectronic Integrated Circuits Conference, Paper 33, SPIE International Symposium on Optoelectronics'97, Vol. 3006, pp. 272–279, San Jose, CA, February 8–14, 1997.
48. S. Sakamoto, R. Spickermann, and N. Dagli, "Novel Narrow Gap Coplanar Slow Wave Electrode for Traveling Wave Electrooptic Modulators," Electron. Lett., 31, 1183–1185, 1995.
49. R. Spickermann, S. R. Sakamoto, and N. Dagli, "In Traveling Wave Modulators Which Velocity to Match?" Proceedings of IEEE/LEOS'96, 9th Annual Meeting, Paper WM3, Boston, MA, November 18–21, 1996, pp. 97–98.
50. W. W. Rigrod and I. P. Kaminow, "Wide-Band Microwave Light Modulation," Proc. IEEE, 51, 137–140, 1963.
51. S. R. Sakamoto, C. Ozturk, Y. T. Byun, J. Ko, and N. Dagli, "Low Loss Substrate-Removed (SURE) Optical Waveguides in GaAs/AlGaAs Epitaxial Layers Embedded in Organic Polymers," IEEE Photonics Technol. Lett., 10, 985–987, 1998.
52. S. R. Sakamoto, A. Jackson, and N. Dagli, "Substrate Removed GaAs/AlGaAs Modulator," IEEE Photonics Technol. Lett., 11, 1244–1246, 1999.
53. J. H. Shin, S. Wu, and N. Dagli, "Substrate removed low drive voltage GaAs/AlGaAs semiconductor electro-optic phase modulators," Integrated Photonics Research Conference Proceedings, Paper IThD-3, San Francisco, CA, June 30–July 2, 2004.
54. J. H. Shin, S. Wu, and N. Dagli, "High-Speed, Low-Voltage Substrate-Removed GaAs/AlGaAs Electro-optic Modulators," International Topical Meeting on Microwave Photonics (MWP '04), Paper W2-5, Ogunquit, Maine, October 4–6, 2004.

5 High-Speed Polymer Optical Modulators and Their Applications

Harold R. Fetterman and William H. Steier

CONTENTS

5.1 INTRODUCTION

Polymer electro-optic devices have been extensively studied and explored during the last few years. They have intrinsic advantages over conventional materials such as high-speed operation, compatibility with other materials and substrates, and the ability to make complex configurations and arrays. On the other hand, there have been problems with thermal stability, power handling, and drift. These problems are currently the subject of intense research and must be solved before commercial devices become available.

This chapter will present the latest developments in high polymer device technology and discuss a number of examples of the use of new material combinations. New transitions from the active materials to low-loss guide structures have been developed. As part of the discussion of designs, we will present the use of microresonators and flexible substrates. The discussion will then shift to novel configurations. Examples will be presented that show how future polymer devices will have applications that are very difficult, if not impossible, with traditional materials.

Examples will include new types of DOS switches, linearized modulators, active microwave phase shifters, and active frequency shifters. The use of arrays and applications to high-speed DSP are just beginning and they will be briefly presented. In addition to the current devices and applications, the new integrated systems have drivers and filters on the same substrates. This will lower the cost of these components dramatically and will lead to greatly expanded roles in communications, interconnections, and processing.

5.2 ELECTRO-OPTIC TRAVELING WAVE MODULATORS

The advantages of polymers for high-bandwidth electro-optic modulators have been recognized for several years [1,2] and there have been recent significant advances in modulators made using the guest-host polymer CLD/APC [3,4,5]. The properties of this polymer, the fabrication procedure, and the properties of modulators using this material have been reported in detail [3]. The polymer has an r_{33} of 65 pm/V at 1300 nm and 55 pm/V at 1550 nm, measured in single thin films. Straight buried ridge waveguides fabricated by RIE have a loss of 1.2 dB/cm at 1550 nm for both the TM and TE. Push-pull poled devices in which the arms of the interferometer are poled in opposite directions have the lowest V. The measured low frequency V of the push-pull devices using the CLD-75 chromophore and a 2-cm interaction length is 1.2 V at 1300 nm and 1.8 V at 1550 nm. The r_{33} achieved in the devices is ~70% of that achieved in the single films. The chip loss for the 3-cm long device (2-cm interaction length) is 4–5 dB, which is consistent with the waveguide loss measurements. The contrast ratio in all devices was > 20 dB. Figure 5.1 shows the fabrication procedure and dimensions of a typical device and the dual-arm microstrip line that is used in the push-pull devices [6]. Figure 5.2 gives the fabrication procedure. The EO polymer core of the waveguide is electrode poled in a nitrogen atmosphere to prevent chromophore damage. The various cladding materials and their properties are given in [3]. Using the CLD chromophore in a PMMA host, Shi has reported a V of 0.8V @1300 nm in a 3-cm long modulator [7].

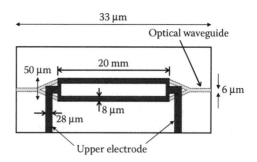

FIGURE 5.1 Layout of a dual-arm push-pull Mach-Zehnder modulator.

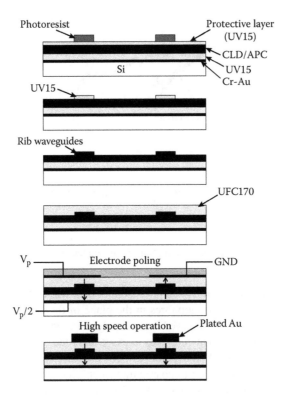

FIGURE 5.2 Fabrication procedure of a typical etched rib waveguide dual-arm Mach-Zehnder modulator. Typical dimensions are as follows: 2.5 um thick lower and upper cladding layers, 3.0 um thick core layer, 0.8 um rib height, 4 um rib width.

Because of low dispersion of polymers, the velocity mismatch between the optical wave and the rf modulation wave is not a factor in determining the bandwidth of polymer modulators up to modulation frequencies of over 100 GHz. Since a device design that corrects for a velocity mismatch [8] is not required, the polymer modulators are therefore of relatively simple design and have a very good overlap between the rf and optical fields. The polymer devices use a microstrip line for the microwave signal, in contrast to the $LiNbO_3$ devices, which typically use a coplanar transmission line. The high-frequency response of the polymer modulators is determined by the microwave loss of the microstrip lines. The 3-dB bandwidth is defined as the frequency at which the detected rf power falls to half of the low frequency value or when the effective V of the modulator increases by $\sqrt{2}$. This is also the frequency at which the S_{21} of the transmission line becomes −6.34 dB. Since the rf loss of most polymers is low [9], the rf loss of the microstrip line is dominated by the metal loss. The polymer modulators we have fabricated using Au electrodes have an rf loss coefficient of ~0.7 dB cm^{-1} $GHz^{-1/2}$ [4]. A 2-cm interaction length device has an rf bandwidth of ~17 GHz, and a 1-cm device has an rf bandwidth of ~35 GHz. Figure 5.3 shows the rf response of the 1- and 2-cm devices. Lower loss coplanar lines have been reported [10] and therefore one should expect a coplanar

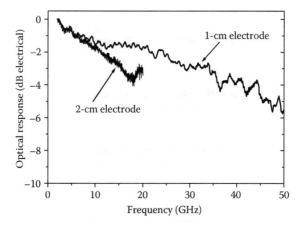

FIGURE 5.3 RF response of 1 and 2 cm long Mach-Zehnder modulators.

design of the polymer modulators will have a larger bandwidth. The coplanar devices will require poling of the chromophores in the plane of the film [11].

There is a bias instability observed in most polymer modulators. When a DC bias voltage is applied, the bias point continues to move and requires a continually higher-bias voltage to stay at the quadrature point [12]. The presence of this instability depends on the polymers used and is due to charge movement that shields the core from the applied bias voltage. In some modulators this drift can be stabilized and the bias point tracked by a bias control system [13]. A very effective and stable biasing has been reported using the thermo-optic effect [14]. One arm of the Mach-Zehnder interferometer is heated by an electrode and because of the relatively large thermo-optic effect in polymers ($dn/dT \sim 10^{-4}$), only a small electrical power is required for quadrature biasing. The thermal approach is stable and essentially drift free.

The long-term thermal stability and photostability of polymer devices are serious system issues. While these questions have not been completely resolved, a good understanding of the causes has evolved and promising progress has been made. The long-term thermal stability is related to the thermal stability of the poling of the polymer. To achieve the EO effect, the chromophores in the polymer must be aligned to a high degree; this is conventionally done by applying a poling electric field at a high temperature. If the modulator is operated at an elevated temperature over a length of time, this alignment will relax and the EO coefficient will decrease. The key to maintaining the alignment and improving the thermal stability is therefore to harden the polymer host either by using a high-T_g material or by cross linking the polymer after the poling is complete [15,16,17]. Typical system applications require long term stability at 85°C. Modulators fabricated with the CLD/APC material show a 25% increase in V after 40 days at 60°C in air. There is research underway to improve the thermal stability. An EO dendrimer material with cross-linking which shows only a 10% decrease in the EO coefficient for 100 hours at 85°C in air has been reported [18] and a side-chain-linked polymer has also shown stable high-temperature operation.

The photostability is related to the relatively high optical intensity in the waveguides of a typical modulator even at input powers of 10s of milliwatts. The peak absorption of the chromophores occurs near 750 nm but at these high intensities, the 1300- or 1550-nm optical power can damage and bleach the chromophores over a long time period. The result of this damage is a decrease in the index of refraction and a decrease in r_{33} [19,20,21]. This optical damage requires the presence of atomic oxygen, and the key to preventing the damage is to exclude oxygen. This can be done by packaging, by using denser host polymers that inhibit the diffusion of oxygen [22], or by including moieties in the polymer which attach the oxygen. Figure 5.4 shows the change in insertion loss and V of a modulator that is packaged in the inert gas Ar. The increase in the insertion loss when the modulator is operated in air is due to the decrease in the index of refraction of the core and the subsequent loss of waveguiding near the input. After packaging, an inert gas is flowed through the package for ~30 min which allows the oxygen trapped in the polymer during processing to diffuse out. The inert gas is then sealed into the package.

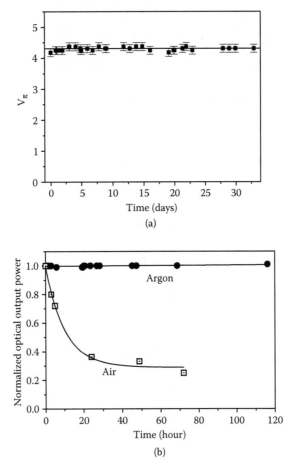

FIGURE 5.4 *V* and insertion loss as a function of time.

5.3 MICRORESONATOR MODULATORS

An example of an interesting structure that can be made using these polymers is a ring-type resonator. Waveguide microring resonators have been used for narrow band wavelength filtering and, if an EO material is used for the ring waveguide, can be used as low-voltage modulators or switches [23]. A typical ring resonator coupled to input and output waveguides and the concept of modulation using the rings are shown in Figure 5.5. The diameter of the ring is limited by the bending or radiation loss, and a small ring radius requires a large index difference between the core and the cladding. The maximum index difference possible in polymers is ~0.3 and the minimum ring diameter is ~50 µm.

If a modulating field, E_z, is applied, the change in index for the TM polarization is

$$\Delta n_z = \frac{1}{2} n^3 r_{33} E_z. \tag{5.1}$$

This change in index can be used as an EO modulator if the operating wavelength is near the half-power point as shown in Figure 5.5.

Electro-optic modulation for TM polarization in microresonators fabricated in unstrained cubic III-V semiconductors such as GaAs and GaP is not possible because of the crystal symmetry [24]. In these materials, for an applied E_z, the average index around the ring changes only for TE polarization.

For the resonator coupled to a single waveguide, the through port transmission can be written as

$$\frac{I_3}{I_1} = \frac{a^2 - 2ra\cos(\phi) + r^2}{1 - 2ra\cos(\phi) + r^2 a^2}, \tag{5.2}$$

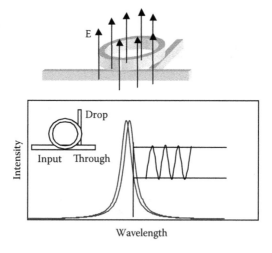

FIGURE 5.5 Light modulation concept using a single MR.

where

a = E field transmission coefficient for one round trip in the ring,

r = E field reflection coefficient of the coupling from the waveguide to the ring, $\phi = \frac{2\pi n_{eff} R}{\lambda}$,

n_{eff} = the effective index of refraction of the mode in the ring waveguide,

R = radius of the ring,

I_3 = intensity out of the coupling waveguide, and

I_1 = intensity into the coupling waveguide

The resonant frequency, $(\omega_0 = 0)$ occurs when $a = r$. Assuming critical coupling, the maximum sensitivity of the modulator or the point where the slope is maximum occurs at a detuning given by

$$(\omega - \omega_0) = \frac{\sqrt{3}}{3} \frac{\Delta\omega_{FWHM}}{2} \tag{5.3}$$

where $\Delta\omega_{FWHM}$ = full width at half maximum of the resonance. At this point the modulator is biased at

$$\left. \frac{I_3}{I_1} \right|_{(\omega-\omega_0)=\frac{\sqrt{3}}{3}\frac{\Delta\omega}{2}} = \frac{1}{4}. \tag{5.4}$$

The sensitivity, S, of the modulator can be written as

$$S = \frac{\partial\left(\frac{I_3}{I_1}\right)}{\partial V_z} = \frac{3\sqrt{3}QKn_{eff}^2 r_{33}}{8d}, \tag{5.5}$$

where $E_z = \frac{V_z}{d}$, K is the confinement factor that expresses the overlap of the optical field, the EO polymer, and the rf field, and Q is the quality factor of the resonator. The higher-Q device will have higher sensitivity, as expected.

In traveling-wave electro-optic polymer modulators, the bandwidth is usually determined by the rf loss of the microstrip line, as noted in the earlier sections. Since the microresonator modulator is usually small compared to the modulation wavelength, the rf electrode loss does not determine the bandwidth, but rather the bandwidth is set by the optical response of the microresonator. For a given transfer function $H(i\omega)$, the group delay is given by

$$\tau_g = \frac{\partial}{\partial\omega}(\arg(H(\omega))). \tag{5.6}$$

In this equation $arg(H(\omega))$ is the phase response of the resonator and the group delay is basically the time it takes for a photon to pass through the resonator. This can be considered as the time the photons spend in the resonator. For a single pole resonator,

$$H(\omega) = \frac{i\,\frac{\omega-\omega_0}{\Delta\omega_{FWHM}/2}}{1+i\,\frac{\omega-\omega_0}{\Delta\omega_{FWHM}/2}} \qquad (5.7)$$

and

$$\tau_g = \frac{\frac{1}{\Delta\omega_{FWHM}/2}}{1+\frac{(\omega-\omega_0)^2}{\Delta\omega_{FWHM}^2/4}}. \qquad (5.8)$$

At the maximum sensitivity point, this group delay is given by:

$$\tau_g = \frac{3}{2\Delta\omega_{FWHM}} \qquad (5.9)$$

Since the electrode is small, we can consider it as a lumped circuit and therefore the photon will experience the integrated average of the electric field over the time, τ_g, it spends in the resonator. The average value for a sinusoidal electrical signal can be written as

$$\bar{V} = \frac{1}{\tau_g}\int_t^{t+\tau_g} V\sin(\omega_{rf}t)dt \approx \frac{\sin(\omega_{rf}\tau_g)}{\omega_{rf}\tau_g}V\sin(\omega_{rf}t), \qquad (5.10)$$

where ω_{rf} is the modulation frequency. The 3-dB electrical bandwidth, f_{3dBe}, of an EO modulator is defined as the frequency at which the received rf power drops to 0.5 of the low-frequency value or when the effective voltage on the modulator drops to 0.707 of the low-frequency value. By setting the sine function in Equation 5.10 to 0.707 and using Equation 5.9, we find:

$$f_{3dBe} \approx \frac{\Delta\omega_{FWHM}}{2\pi} = \Delta\nu_{FWHM} = \frac{c}{\lambda Q}. \qquad (5.11)$$

The bandwidth of the modulator is inversely proportional to the Q. However, the product of the sensitivity and the bandwidth is not related to the Q and is given by

$$f_{rf(3dB)}S = 0.65\frac{n^2 r_{33}K}{d}\frac{c}{\lambda}. \qquad (5.12)$$

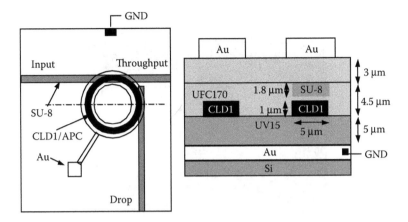

FIGURE 5.6 Large diameter electro-optic microresonators.

One can improve the sensitivity but at the cost of reduced bandwidth. The only way to improve the sensitivity (keeping bandwidth constant) is to increase the electro-optic coefficient. This fundamental trade-off was recently pointed out by Williamson [25] and by Gheorma et al. [26].

The schematic of the microring modulator is shown in Figure 5.6. The same polymers and fabrication procedures were used in these modulators as in the traveling wave modulators. The index difference between the CLD/APC core and the UFC170 cladding is 0.04 and this limits the ring diameter to larger than 300 μm for 1550 nm and 200 μm for 1300 nm. The effective refractive index of the SU-8 waveguide matches the effective refractive index of the mode of the CLD1/APC microring.

Cleaved fiber was butt-coupled to the input and output waveguides and a tunable New Focus laser at 1300 nm was used as the source. Figure 5.7(a) shows the drop port power as a function of wavelength for a 1500 μm diameter device. As can be seen from this figure, Δv_{FWHM} @ 1300nm is ~3 GHz for TM polarization (~4 GHz for TE). Hence the Q is 7.6×10^4 and 6.2×10^4 for TM and TE, respectively. The Q

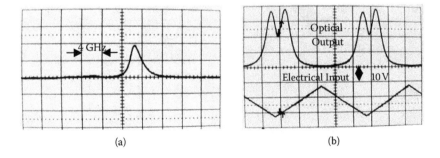

FIGURE 5.7 (a) Measured dropped power at zero applied voltage as a function of wavelength. (b) Modulated dropped power and the applied voltage on the electrode at 1300 nm for a 1500 ∝m diameter resonator at a fixed laser wavelength.

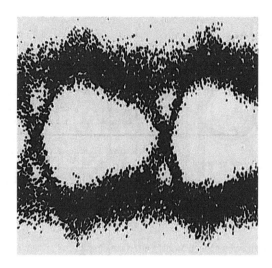

FIGURE 5.8 Eye diagram for data transmission at 1 Gb/sec using electro-optic MR.

values at 1500 nm are approximately half the size of the values at 1300 nm. Figure 5.7(b) shows the modulated TM light at the drop port of the device when a saw-tooth voltage is applied to the electrode. One figure of merit for the EO microresonator is the voltage, V_{FWHM}, required to shift the resonant frequency by $\Delta\upsilon_{FWHM}$. This is measured to be 4.86 V @ 1300 nm and this corresponds to an r_{33} of 33 pm/V. The typical r_{33} of CLD1/APC in Mach-Zehnder modulators is 50 pm/V @ 1300 nm and is probably less in the microresonators because of less efficient poling. V_{FWHM} at 1550 nm for this device is 16 V. For a smaller (600 μm diameter) device the Q is slightly lower (4.7×10^4 and 5.8×10^4 for TE and TM, respectively) and the V_{FWHM} is 9 V. The sensitivity of the 1300-nm modulator when biased to the maximum slope is ~ 0.3 V^{-1}, and the sensitivity- bandwidth product is ~ 0.9 GHz/V.

To demonstrate digital modulation, a 1-Gb/sec, 1-V data stream was applied to the electrode of the 1500-mm diameter ring. The relatively clear eye diagram is shown in Figure 5.8.

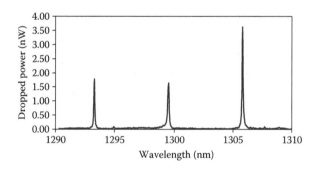

FIGURE 5.9 Response of the 50-μm diameter electro-optic modulator as a function of laser wavelength.

TABLE 5.1
The Performance of Three Different Devices at 1300 nm

Device Diameter (µm)	FSR (nm)	Δv_{FWHM} (GHz)	V_{FWHM} (V)	Tuning (GHz/V)
150	1.7	12	10	1.2
70	4.3	16	20	0.8
50	6.2	18	36	0.5

One potential application of microresonator modulators is in WDM communication systems where each wavelength channel is modulated by a microresonant modulator tuned to that wavelength. The number of WDM channels will be partly determined by the free spectral range (FSR) of the microresonators. To increase the FSR, we have fabricated a series of smaller devices by using Teflon as the cladding material. The fabrication is similar to that of the larger devices. Figure 5.9 shows the wavelength scan of the 50-mm diameter ring at 1300 nm; the FSR is 6.2 nm. Table 5.1 summarizes the properties of the high FSR modulators.

5.4 FLEXIBLE LOW-VOLTAGE ELECTRO-OPTIC POLYMER MODULATORS

Perhaps the most unique feature of the EO polymer devices is their compatibility with a variety of substrates such as Si, GaAs, or plastic [27,28] This has made it possible to integrate polymer modulators with prefabricated high-speed electronics [28] and with other photonic materials. A plastic substrate can be used to implement a flexible guided-wave high-speed optical modulator, a device that is not possible using crystalline EO materials. The flexible EO waveguide device offers the possibility of conforming to a curved surface such as an antenna or an aircraft surface. Such a device also could apply to large optically controlled phased-array antennas for space that are unrolled after the satellite is in orbit. The flexible polymer modulator also shows the robustness of the technology.

In an earlier work, poled EO polymer films were lifted from a substrate using a water-soluble layer [29,30]. Mylar was used as a substrate for a flexible modulator that was inserted into a W-band rectangular microwave waveguide to demonstrate a traveling-wave modulator at 95 GHz [31]. The modulator was released from the fabrication substrate using a solvent to dissolve the adhesive. The limitation with both of these approaches is the difficulty in finding an adhesive to attach the flexible substrate to the fabrication substrate that can later be dissolved and that can withstand the temperatures used in poling the polymer film. This has become a larger problem with the high-T_g polymers that are poled at higher temperatures and are stable at high temperatures.

Shi et al. [32] has reported on a dry lift-off technique that is based on the poor adhesion of a gold-glass interface and that can be used at high temperatures. We have developed an approach using a polymer, SU-8, that is based on the difference between the adhesion of the polymer to gold and to an Si surface, and that meets the temperature requirements.

The basic approach to make the flexible device is to lift off a multiple-layer polymer film from a Si substrate. The polymer film consists of a release layer of SU-8, a polymer substrate layer of UV-15, and cladding-core-cladding polymer modulator layers.

SU-8, a commercial photo-resist, has a striking difference between adhesion on a Si wafer and adhesion on a gold metal surface. We found that the SU-8 layer coated on an Au surface has very poor adhesion and it is therefore possible to use the Au/SU-8 interface as the release interface for a lift-off process. On the other hand, SU-8 has good adhesion on a Si surface. Hence, one can lift off selected areas by patterning the Au coating on an Si substrate.

Thick films of the SU-8 polymer are possible and this material might serve as the substrate for the flexible device. However, after lift-off, the baked thick films of SU-8 are relatively hard and cannot be bent without cracking; they often crack during dicing. To overcome this problem without losing the unique adhesion property, we make use of two polymer layers for the flexible substrate. The 3-μm SU-8 layer was coated on an Si substrate patterned with an Au metal layer. Then a UV curable epoxy, UV-15, was coated and cured in several layers until the total thickness reached 100 μm. UV-15 has been used as the cladding material in our laboratory and has enough thermal stability to withstand poling temperatures over 150°C [33]. On the top of this layer, the EO polymer modulator was fabricated using our standard fabrication procedure [33]. The bottom electrode was formed by vacuum evaporation of Cr and Au with thicknesses of 100 Å and 500 Å, respectively. UV-15 was spin coated to have a thickness of 3 μm as the lower cladding. A guest-host EO polymer, CLD-1/APC, was used as the 2.2-μm core layer. Mach-Zehnder waveguide patterns with a waveguide width of 4 μm were formed by a standard photolithography. Reactive ion etching (RIE) in oxygen was carried out to form a rib height of 0.6 μm. The 3.2-μ*m* upper cladding is UFC170A. Au poling electrodes of 500 Å with a length of 20*mm* were fabricated on the cladding by the vacuum evaporation, standard photolithography, and wet etching. For the push-pull operation, two arms of the Mach-Zehnder modulator were poled in opposite directions. Each arm was poled by applying a voltage of 400 V at 145°C in nitrogen. Finally, the edge of the wafer was diced and the modulator lifted off at the SU-8/Au interface. The fabrication procedure is schematically shown in Figure 5.10. Figure 5.11 shows the photograph of the fabricated EO polymer devices and shows the degree of flexibility. Because of the difference in adhesion of the SU-8, we were able to leave Si pads on each end of the flexible modulator that made fiber pigtailing and handling easier.

To measure the performance of the modulator, TM polarized light of 1550 nm was coupled to the device through the single-mode fiber. The output light was collected by an objective lens and focused onto a photodetector. The half-wave voltage *V* and the extinction ratio were measured by applying a 1-KHz signal with

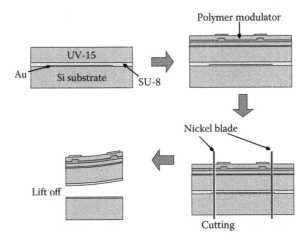

FIGURE 5.10 Fabrication procedure for the EO polymer modulator on the flexible substrate.

a triangular waveform. As shown in Figure 5.12, the measured V was 2.6 V, which corresponds to r_{33} of about 30 pm/V, and the extinction ratio of the modulator was better than 20 dB. These results were essentially the same as those we obtained from the EO polymer device on a conventional Si substrate fabricated by the same process. Therefore, we could verify that there was no degradation in the device performance when the EO polymer devices were fabricated on the flexible substrate.

One of the major questions concerning flexible devices is how much they can be bent before their performance is affected. The insertion loss may depend on the bending radius if the waveguides begin to radiate into the substrate or if the material cracks. In addition, the stresses induced by the bending may change V, the extinction ratio, and the bias point.

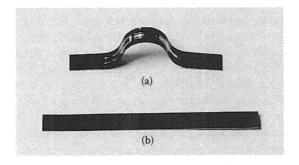

FIGURE 5.11 Photograph of the EO polymer devices on the flexible substrate. (a) Flexed EO polymer device on the flexible substrate. (b) Straight EO polymer device on the flexible substrate.

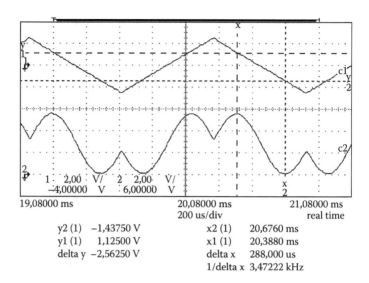

| 1 2,00 V/   2 2,00 V/ | | |
| -4,00000 V   6,00000 V | | |

| 19,08000 ms | 20,08000 ms | 21,08000 ms |
| | 200 us/div | real time |

y2 (1)	-1,43750 V	x2 (1)	20,6760 ms
y1 (1)	1,12500 V	x1 (1)	20,3880 ms
delta y	-2,56250 V	delta x	288,000 us
		1/delta x	3,47222 kHz

FIGURE 5.12 Oscilloscope trace of the input and output signal during V measurement. The result shows that the MZ modulator on the flexible substrate has 2.6 V of half-wave voltage at 1550 nm.

To measure the change of the insertion loss, we fabricated flexible straight waveguides using the same procedure as the modulators and pigtailed them to single-mode fiber. The flexible waveguides were then bent around a cylinder of radius R and the insertion loss was measured [34] as shown in Figure 5.13(a). Figure 5.13(b) shows the insertion loss dependence on the bending radius, R. Because of the large refractive index difference between the core and cladding (~ 7 %), the insertion loss is not increased until the bending radius is reduced to 1.5 mm. After repeated bending of the waveguides, no cracks were seen and no change in the insertion loss was measured.

To measure the changes in V, extinction ratio, and bias point, a flexible modulator was fiber pigtailed on both ends using epoxy and quartz plates to support the fiber. The 35-mm long middle part between the plates remained flexible. The quartz plates were then mounted on independent stages and the modulator flexed by moving the plates together. The modulator flexed in the shape shown in Figure 5.11(a), where the bending radius can be calculated from the amount of plate movement by an S-bend approximation. V and the extinction ratio were unchanged within our measurement errors at a bending radius of 5 mm, the smallest we could obtain in the experiment.

We did, however, observe a repeatable shift in the modulator bias point as a function of the bending radius, as shown in Figure 5.14. There is no DC bias voltage on the modulator, so the bias point is set by the path length difference in the two arms of the Mach-Zehnder. We believe the bias point shift is due to the stresses induced by the bending. The push-pull poled modulators typically are not balanced because the poling in each arm is not identical. In this unbalanced interferometer, equal stress in both arms will shift the bias point.

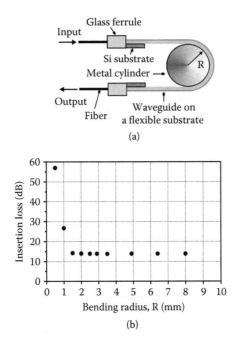

(a)

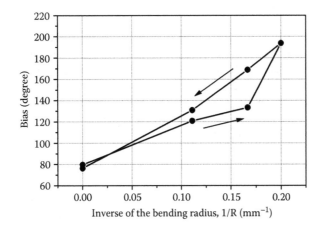

(b)

FIGURE 5.13 Insertion loss measurement of the EO polymer device on the flexible substrate. (a) Measurement method. (b) Insertion loss change depending on the bending radius. The insertion loss remains unchanged with a bending radius of 1.5 mm.

FIGURE 5.14 Bias point shift depending on the bending radius. The modulator bias point was shifted in the direction of the arrow when the flexible modulator was bent and released.

We have demonstrated a method to fabricate flexible polymer devices based on the unique property of SU-8 that can be easily peeled off from an Au surface, although it adheres very well on silicon substrate. The EO polymer modulators formed on a flexible polymer substrate have essentially the same half-wave voltage (2.6 V) and the same extinction ratio (>20 dB) as the conventional polymer modulators fabricated on Si substrates. There appears to be no significant change in the insertion loss, V, or extinction ratio for bending radii of a few millimeters. The minimum bending radius appears to be much smaller than the bending radii that would be needed in most applications. The difference in the adhesion of the SU-8 to Au and to Si opens some interesting device fabrication possibilities for lifting off patterned shapes or lifting off only selected areas.

5.5 HIGH-SPEED LINEARIZED POLYMERIC DIRECTIONAL COUPLER MODULATOR (LPDCM)

One of the major problems with modulators made with these electro-optic materials is dynamic range and linearity. Electro-optic intensity modulators, such as Mach-Zehnder modulators, have inherently nonlinear transfer functions. Harmonics and intermodulation distortions due to this nonlinearity will degrade the system performance and decrease the link dynamic range [35]. Various linearization approaches have been proposed, including dual-polarization techniques [36], parallel or series cascaded configurations [37,38], and electronic predistortion schemes [39]. Most schemes involve multiple modulators, which represent a significant implementation challenge. Thus, the need for a simple design and minimum coordination among inputs is one of the most crucial considerations when it comes to practical implementation.

Based on a two-section directional coupler modulator configuration, a linearized directional coupler modulator based on EO polymer has been fabricated and tested. EO polymer has shown its remarkable potential in high-speed application since it has many intrinsic advantages such as high electro-optic coefficients, less phase velocity mismatch, and flexibility in fabrication. This modulator, consisting of only a simple uniform electrode configuration, requires only a single driving source and hence is suitable for high-speed applications. In addition, as a Y-fed structure, it is known to benefit from the built-in quadrature point, which eliminates the necessity of extra biasing voltages. Therefore, this linearized Y-fed polymeric directional coupler modulator is very promising for high-frequency, broadband operation.

Figure 5.15 shows a two-section directional coupler modulator with two electrode sections of opposite polarity. A matrix representation is used to evaluate the transfer function of this two-section directional coupler [40], shown as the following form:

$$
\begin{bmatrix} R_{out} \\ S_{out} \end{bmatrix} = \mathbf{BA} \begin{bmatrix} R_{in} \\ S_{in} \end{bmatrix} = \begin{bmatrix} b_{11} & b_{12} \\ b_{12} & b_{11}^* \end{bmatrix} \begin{bmatrix} a_{11} & a_{12} \\ a_{12} & a_{11}^* \end{bmatrix} \begin{bmatrix} R_{in} \\ S_{in} \end{bmatrix}, \quad (5.13)
$$

where R_{in} and R_{out} are the input and output optical fields. Matrices A and B represent the first and the second section of the directional coupler, respectively, and

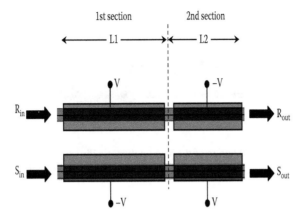

FIGURE 5.15 Schematic of the two-section directional coupler.

their coefficients are given by the following expressions based on the couple-mode theory:

$$a_{11} = \cos\left(\kappa L_1 \sqrt{1+x^2}\right) - j\frac{x}{\sqrt{1+x^2}}\sin\left(\kappa L_1 \sqrt{1+x^2}\right),$$

$$a_{12} = -j\frac{1}{\sqrt{1+x^2}}\sin\left(\kappa L_1 \sqrt{1+x^2}\right),$$

$$b_{11} = \cos\left(\kappa L_2 \sqrt{1+x^2}\right) + j\frac{x}{\sqrt{1+x^2}}\sin\left(\kappa L_2 \sqrt{1+x^2}\right),$$

$$b_{12} = -j\frac{1}{\sqrt{1+x^2}}\sin\left(\kappa L_2 \sqrt{1+x^2}\right),$$

where L_1 and L_2 are the lengths of the first and the second coupling section. Here, x is the normalized driving voltage $\Delta\beta(V)/2$, where $\Delta\beta(V)$ is the mismatch between the propagation constants of the coupled waveguides introduced by the applied voltage V. κ is the coupling coefficient which, in a normal directional coupler with two identical waveguides, is determined by $\kappa = 2/l$. l is defined as the coupling length, where 100% power is transferred from one arm to other.

From Equation 5.13, it is clear that the outputs fields of the two-section directional coupler are determined by the parameters L_1 and L_2. Tetsuya Kishino et al. [41] already proves theoretically that the linearity of the transfer curve in a two-section directional coupler modulator can be increased significantly by appropriately selecting the lengths, i.e., L_1 and L_2. Specifically, if l denotes one coupling length

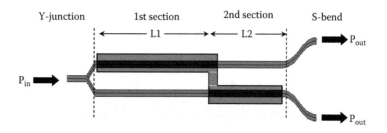

FIGURE 5.16 Schematic diagram of the Y-fed polymeric directional coupler modulator. The waveguide width is 4 μm and the separation between two waveguides is 14 μm.

of the passive waveguide as mentioned before, the optimized length for both sections, determined as $L_1 = 2.3\,l$ and $L_2 = 1.05\,l$, gives the most efficient suppression in third-order distortion.

The structure of the Y-fed linearized polymeric directional coupler modulator (LPDCM) is shown in Figure 5.16. This schematic is not to scale for illustration purposes. The structure consists of a single-mode input waveguide feeding a symmetric Y-junction which branches to a two-section directional coupler. Input light is coupled into a straight waveguide and then evenly splits into two arms after propagating through the Y-junction, i.e., $R_{in} = S_{in} = \frac{1}{\sqrt{2}}$ in Equation 5.13 for this case. The outputs of the coupled-waveguide section are then shaped as S-bends for coupling to the output single-mode fiber. Since the waveguide structure is symmetric, both outputs have the same power for no applied voltage, which means this device is already passively biased to the quadrature point. Based on the theory described earlier, in order to obtain the most linear performance, L_1 and L_2 are set to be $2.3l$ and $1.05l$, respectively. The dual-electrode structure, like that shown in Figure 5.15, has been modified to single-electrode, as shown in Figure 5.16. This design can simplify coupling to a single microwave source, which is exceptionally suitable for high frequency operation.

The experimental transfer function of the fabricated LPDCM is shown in Figure 5.17. The solid line represents the experimental transfer function of LPDCM and is compared with the normal Mach-Zehnder modulator. The transfer function of the linearized DCM shows an enhanced improvement in linearity. In terms of modulation depth, the transfer curve of the linearized DCM also has a wider linear region for modulation.

In analog links, one of the methods to evaluate the linearity performance of a modulator is to apply a two-tone test and calculate the dynamic range [42]. Based on the experimental transfer function, a two-tone analysis is performed. Two electrical test signals, with equal amplitude sinusoidal modulations at frequencies f_1 and f_2, are used to drive the modulator. The output power of the third-order intermodulation terms, $2f_1 - f_2$ and $2f_2 - f_1$, are evaluated and plotted in terms of input power per channel. The same analysis is also applied to the Mach-Zehnder interferometer and shown in Figure 5.18.

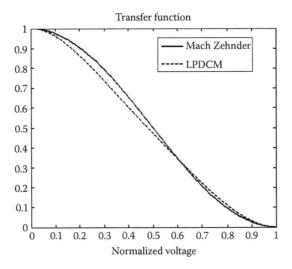

FIGURE 5.17 The experimental transfer function of an LPDCM compared with that of a Mach-Zehnder modulator.

Figure 5.18 shows the output RF signal power and third-order intermodulation power as a function of the input signal power for a fiber-optic link. The dynamic range of MZM is 109.9 dB for a 1-Hz bandwidth biased at the quadrature point. For LPDCM in the same photonic link, the dynamic range is 119.4 dB, which shows a 10-dB enhancement in dynamic range.

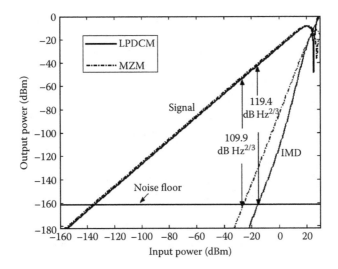

FIGURE 5.18 The signal and third-order intermodulation distortion (IMD) of an LPDCM and a Mach-Zehnder modulator.

5.6 ELECTRO-OPTIC POLYMERIC DIGITAL OPTICAL SWITCHES (DOSS)

Other devices are required for modern systems that have a switch-like behavior. High-speed and high-performance optical switches are vital for optical communication networks and optical signal processing systems. Among those, the optical switches based on the electro-optic (EO) effect are promising for applications that require high-speed switching such as optical burst switching and optical packet switching. Mach-Zehnder interferometer (MZI) switches and directional coupler (DC) switches have been under intensive investigation for a long time, but they both have very limited fabrication tolerance [43]. Furthermore, bias control networks and precise switching voltages are required in MZI switches and DC switches to achieve high performance.

Digital optical switches (DOSs) were first proposed in 1987 based on the principle of mode evolution (or, equivalently, adiabatic propagation) [44]. These types of devices are known for their good fabrication tolerance and insensitivity to wavelength and polarization. There is no need for bias control of DOSs, and their output has been shown to have a steplike response to applied driving voltages, which is quite desirable for the simplification of control electronics in large-scale optical cross connects.

Because of the large electro-optic coefficient of polymer materials, refractive index changes can be realized by lower driving voltages than in other EO materials. Since there is better optical/RF velocity matching inside the EO polymer, switching at high speeds can be achieved more easily. Therefore, EO polymers are an excellent choice to realize high-speed DOS devices.

Mode evolution or *adiabatic propagation* (in the following description, we will use these two terms interchangeably) is the basic principle used in DOSs. In guided-wave optics, adiabatic propagation means that the occupation of the optical modes of the system does not change as the waveguide structure changes.

The simplest design for a digital optical switch is an active Y-junction. The waveguide structure and mode evolution of the simplest 1×2 switch is shown in Figure 5.19. The symmetry of the output Y-branch can be broken by applying an external signal (either utilizing a thermal or electro-optic effect) to change the refractive index. Asymmetric waveguide branches can act as mode sorters when designed properly [45]. If the structure change of the branch is gradual enough, the fundamental mode input will stay in the fundamental mode along the whole structure. As a result, most of the power will be gradually directed to the waveguide with higher refractive index. Significant theoretical work [46] has been done studying the mode shapes, mode evolution, and branch shapes in these optical waveguide branches.

The 2×2 DOSs, like that shown in Figure 5.20, are slightly more complicated than 1×2 DOSs. The major difference is that 2×2 DOSs have two input waveguides. Usually, the two waveguides are different and there is a multimode waveguide section that links the input section and the output section. Similar to the output section, the input section is also made to have a gradual structure change.

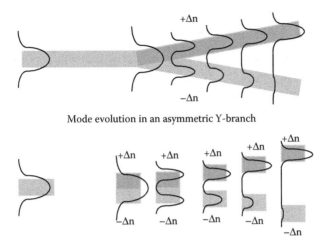

Mode evolution in an asymmetric Y-branch

Local fundamental modes for each section of composite waveguides

FIGURE 5.19 Mode evolution in an asymmetric Y-branch.

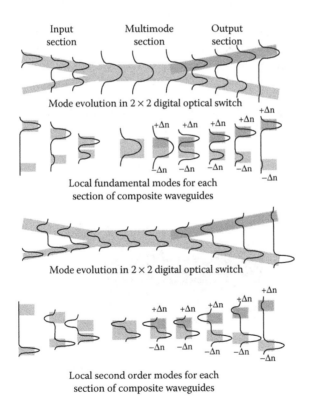

Input section   Multimode section   Output section

Mode evolution in 2 × 2 digital optical switch

Local fundamental modes for each section of composite waveguides

Mode evolution in 2 × 2 digital optical switch

Local second order modes for each section of composite waveguides

FIGURE 5.20 Mode evolution and local modes in a 2 × 2 digital optical switch.

As shown in Figure 5.22, the input from the larger waveguide will excite the fundamental mode in the multimode section and then be directed to the waveguide with higher refractive index. The input from the smaller waveguide will excite the second order mode in the multimode section and then be directed to the waveguide with lower refractive index. By switching the polarity of refractive index change of the output section, we can switch the 2×2 DOS between BAR state and CROSS state. The mode evolution and local modes of different sections of the switch are shown in Figure 5.20.

A ridge waveguide structure is chosen for our DOS, and the structure of the interaction region is shown in Figure 5.21. A waveguide with this structure is fairly easy to fabricate and integrate with other building blocks. The active polymer is APC-CPW1, the lower cladding is UV15LV, and the upper cladding is UFC170A. The cladding materials are both commercially available.

The waveguide layout of our 1×2 EO polymeric DOS is shown in Figure 5.22. Since the DOSs are based on the adiabatic propagation principle, the change of the waveguide structure is kept to a minimum throughout the whole device. The basic structure is an active Y-junction with both arms driven by the voltage applied on the electrodes. The length of the active region is 0.95 cm and the angle between the two arms is around 1.5 mrad to maintain the adiabatic condition along the whole device. Based on our design of the waveguides and the EO coefficient of the polymer material, a change of refractive index of $\Delta n = 0.0001$ is achievable if 10 V is applied to the electrode. By our numerical simulation of the device, this refractive index change is large enough for the DOS to switch. The $\Delta \beta \cdot L$ product for our DOSs is around 3.9, which is quite near to the theoretical prediction of 3.87 for DOSs [47].

After fabrication, the EO polymeric DOSs were tested. The input to the DOS is a piece of conventional single-mode fiber and a $40 \times$ lens is used to collect the optical

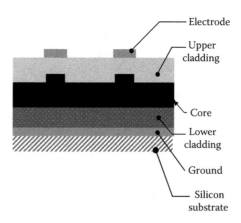

FIGURE 5.21 Cross section of the interaction region of a polymeric DOS.

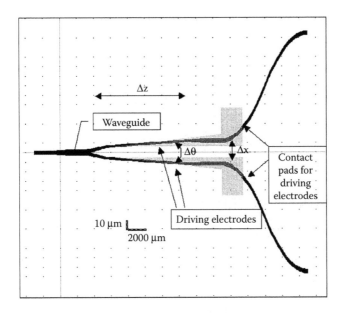

FIGURE 5.22 Waveguide layout and electrode of an EO polymeric DOS (low-speed design).

power from the output of the device. By application of a triangular waveform to the DOS, the modulation curve is measured and shown in Figure 5.23. The device exhibits a steplike modulation curve, which is typical for DOS devices [44,47]. The switching voltage for our DOS is about 14 V. The DOS is designed to have both arms driven by

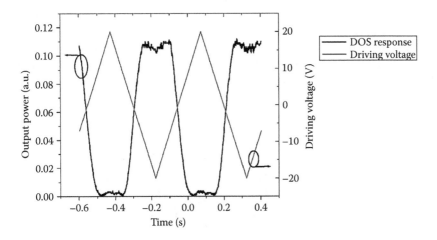

FIGURE 5.23 Driving voltage and response from a DOS.

signals of opposite polarity; however, in this test only one electrode is used for driving, so the equivalent switching voltage is 7 V for this device [48]. This is the lowest switching voltage to our knowledge reported for EO polymeric DOSs and is comparable to DOSs based on LiNbO$_3$ [49]. In this initial demonstration the extinction ratio for the DOS is higher than 20 dB and the insertion loss is around 13 dB.

In summary, electro-optic polymeric digital optical switches (DOSs) have been demonstrated using electro-optic polymers. The devices exhibit low driving voltage (7 V for dual-arm driving), high extinction ratio (20 dB), and steplike response to driving voltage. Promising applications for DOSs include integration into large-scale optical cross connects and for use as high-speed optical burst or optical packet switches.

5.7 MULTIMODE INTERFERENCE (MMI) SWITCHES

Conventional integrated optical switches, such as directional coupler switches or Mach-Zehnder Interferometer (MZI) switches, have very limited fabrication tolerances. Furthermore, in some applications a dc bias is required to obtain a specific phase shift.

In this regard, multimode interference (MMI) devices, which are based on the self-imaging effect [50], have gained considerable popularity in recent years. The self-imaging phenomenon results when an optical field distribution, often in the form of a single-mode access guide, is introduced in a heavily multimode waveguide region. The input field is then coupled into all the modes of the later guide, as shown in Figure 5.24, with varying amplitude coefficients, and the interference of these modes with each other results in multiple "copies" of the original field at various points along the propagation direction. By selecting the appropriate length of the multimode region, we can specify the number of "copies" to be made, which can range from 1 to N. MMI devices can perform as splitters (i.e., by splitting one optical into multiple outputs) or combiners (i.e., by combining multiple input optical signals into a single output) with remarkable efficiency. In addition, MMI devices have large fabrication tolerances and fairly small sizes can be achieved without compromising the efficiency [51,56]. All these features make MMI an ideal candidate for novel devices compared with the directional couplers or Y-branches we have discussed.

Each input waveguide of the MMI coupler generates an object in the input plane of the MMI that, in accordance with the self-imaging theory, creates a set of self images in the output plane. The number of these images, as well as their amplitudes and relative

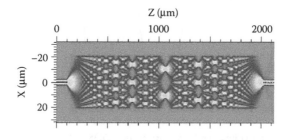

FIGURE 5.24 A 1 × 1 MMI coupler. Note the multiple images formed at intermediate locations.

phases, depends on the length of the MMI section. If output waveguides are placed at the location of these images, input signal can be split into multiple output channels.

The shortest length L_N of an MMI that produces N images of an input field $f_{in}(x)$ of equal amplitudes is given by [52]

$$L_N = \frac{3\pi}{N \cdot \Delta\beta_{01}}, \tag{5.14}$$

$$\Delta\beta_{01} \cong \frac{3\pi^2}{2nk_o W^2}, \tag{5.15}$$

where $\Delta\beta_{01} = \beta_0 - \beta_1$ is the difference between the propagation constants of the two lowest-order modes of the MMI and for a strongly guiding MMI (i.e., with modes almost completely confined in the MMI section) it can be approximated as above. W is the width of the MMI, n is the refractive index, and $k_0 = 2\pi/\lambda_0$ is the propagation constant in vacuum.

The positions, phases, and amplitudes of the N output images are given by the following formulas:

$$f_{out}(x) = \frac{1}{C} \sum_{q=0}^{N-1} f_{in}(x - x_q) \cdot \exp(j\phi_q), \tag{5.16}$$

$$x_q = (2q - N)\frac{W}{N}, \tag{5.17}$$

$$\phi_q = q(N - q)\frac{\pi}{N}, \tag{5.18}$$

$$C = \sqrt{N} \cdot \exp(j\phi_o), \tag{5.19}$$

$$\phi_o = \beta_o L_N + \frac{\pi}{N} + \frac{\pi}{4}(N - 1). \tag{5.20}$$

The amplitude of the outputs is $1/\sqrt{N}$, with a common constant phase φ_0. In addition to this common phase, the relative phase differences between outputs are given by φ_q.

A schematic drawing of a 2×2 nonblocking switch is shown in Figure 5.25. The first MMI evenly splits the optical powers of the two input signals into two output guides. The intermediate electro-optic (EO) phase shifter arms provide active control of the switches [53] and the second MMI works as a combiner of the signals. With proper phase shifts induced through the controlling arms, input signals can be rerouted to the desired output channels. A strictly nonblocking switch should be able to provide all of the possible rerouting configurations. The splitter and combiner are chosen to be identical 2×2 MMIs and all the access guides are single-mode ridge guides.

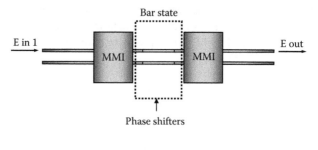

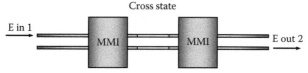

FIGURE 5.25 Diagram of a 2 × 2 nonblocking switch.

The transfer matrix, $M^{(2)}$, for the 2 × 2 MMI can be calculated

$$M^{(2)} = \frac{1}{\sqrt{2}} \cdot \exp(-j.\phi_c^{(2)}) \cdot \begin{bmatrix} 1 & \exp\left(-j \cdot \frac{\pi}{2}\right) \\ \exp\left(-j \cdot \frac{\pi}{2}\right) & 1 \end{bmatrix}, \tag{5.21}$$

$$\phi_c^{(2)} = \beta_0 L_N + \frac{7\pi}{4}, \tag{5.22}$$

where $\phi_c^{(2)}$ is the constant common phase in both output arms. Similarly, the phase shifting section can be represented by a diagonal matrix Φ of the form

$$\Phi = \begin{bmatrix} \exp(j \cdot \phi1) & 0 \\ 0 & \exp(j \cdot \phi2) \end{bmatrix}, \tag{5.23}$$

where $\phi1$ and $\phi2$ are the external phase shifts induced electro-optically in the upper and lower arms, respectively.

The resulting two outputs, E_{out}^1 and E_{out}^2, can now be written as

$$E_{out}^1 = \frac{\exp(-j \cdot \phi_c^{(2)})}{2} \cdot \left\{ (e^{j\phi1} + e^{j\phi2}) \cdot E_{in}^1 + j \cdot (e^{j\phi1} - e^{j\phi2}) \cdot E_{in}^2 \right\} \tag{5.24}$$

$$E_{out}^2 = \frac{\exp(-j \cdot \phi_c^{(2)})}{2} \cdot \left\{ j \cdot (e^{j\phi1} - e^{j\phi2}) \cdot E_{in}^1 + (e^{j\phi1} + e^{j\phi2}) \cdot E_{in}^2 \right\} \tag{5.25}$$

It is easy to see that the condition ($\phi1 = \phi2$) gives the bar state, and the condition ($\phi1 - \phi2 = \pi$) gives the cross state. Instead of the exact phase shifts induced in each arm, only the relative phase difference between the two is of any real significance. We can further reduce the voltage necessary to flip between the bar and the cross state, by a factor of two, by using the push-pull poling scheme for our active polymer.

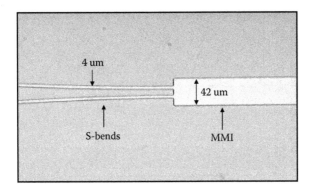

FIGURE 5.26 Photograph of a fabricated 2 × 2 MMI switch.

As the device geometry indicates, the two arms of the phase-shifting segment of the device are supposed to be of equal length. However, even if there are fabrication errors, they will not affect the efficiency of the switch because the active phase shifter can compensate for the relative length variations between the two arms.

We can absorb one of the phase shifts in the control arms, say $\phi1$, in the constant phase $\phi_c^{(2)}$ so that the above equations can now be rearranged as

$$E_{out}^{\ 1} = \frac{\exp(-j \cdot \phi_c^{(2)} + j\phi1)}{2} \cdot \left\{ (1 + e^{j(\Delta\phi)}) \cdot E_{in}^{\ 1} + j \cdot (1 - e^{j(\Delta\phi)}) \cdot E_{in}^{\ 2} \right\}, \quad (5.26)$$

$$E_{out}^{\ 2} = \frac{\exp(-j \cdot \phi_c^{(2)} + j\phi1)}{2} \cdot \left\{ j \cdot (1 - e^{j(\Delta\phi)}) \cdot E_{in}^{\ 1} + (1 + e^{j(\Delta\phi)}) \cdot E_{in}^{\ 2} \right\}, \quad (5.27)$$

$$\Delta\phi = \phi2 - \phi1 . \quad (5.28)$$

It can be seen that only one control arm is sufficient to switch between the bar and the cross state, if it can provide a relative π-degree phase shift (i.e., $\Delta\Phi = \pi$) between the two arms.

We have fabricated both 2 × 2 and 3 × 3 MMI switches in APC/CLD1, as seen in Figure 5.26 and Figure 5.27, respectively. Standard rib waveguides were used for

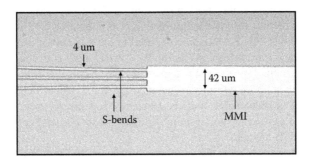

FIGURE 5.27 Photograph of a fabricated 3 × 3 MMI switch.

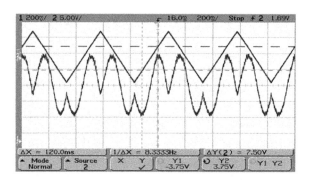

FIGURE 5.28 Response of a 2 × 2 MMI switch indicating a single-arm switching voltage of 7.5 V (3.75 V for push-pull operation).

these devices. Figure 5.28 shows the response of a 2 × 2 MMI switch, indicating that it has a 7.5-V switching voltage. Insertion losses can be reduced in the next generation of devices by fabricating the multimode interference waveguide region of the device as a channel waveguide, which has stronger optical confinement than a standard rib waveguide.

5.8 ELECTRO-OPTIC FREQUENCY SHIFTING USING POLYMER PHASE MODULATORS

Electro-optic (EO) frequency shifting is based on the phase modulation of an optical signal by a properly chosen microwave waveform in an EO modulator. This is a particularly important area of modern optical communications research. To achieve frequency shifting, the modulating waveform must be such that for the given parameters of the modulator (velocity mismatch, RF loss, etc.), the effective phase modulation of the optical signal is a linear ramp in time. As the result of such modulation, the optical signal is frequency shifted by the amount proportional to the slope of this ramp. This process is conceptually similar to the Doppler shift, since in effect they both require linear time dependence of the optical path length.

EO frequency shifting has exciting applications in numerous areas of science and technology. Perhaps the most promising of these applications is the optical wavelength conversion for wavelength-division multiplexing (WDM) communication systems. This method is unique compared to the other all-optical (not requiring light detection and retransmission) methods of wavelength conversion [58]. It does not require additional lasers, and the amount of the shift is controlled by the amplitude of the modulating waveform. Thus, the power of the optical signal is not critical. The conversion is transparent, and its efficiency is close to unity. EO frequency shifting can also be used in spectroscopy, optical gyroscopes, and optical modulators for frequency-shift keying.

Polymer phase modulators are very attractive for use as EO frequency shifters due to high Pockels coefficients and low intrinsic velocity mismatch between optical and microwave signals (with the possibility of tailoring this mismatch to a desired value).

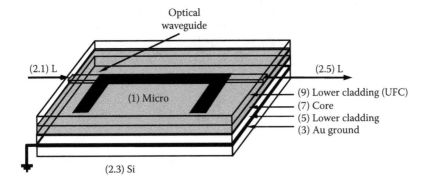

FIGURE 5.29 Schematic diagram of a polymer phase modulator.

A traveling-wave EO phase modulator consists of a nonlinear optical waveguide and a microwave waveguide on top of it (Figure 5.29). The microwave signal changes the refractive index in the optical waveguide, thus phase-modulating light in it. In order to increase the modulation bandwidth, it is desirable to match the velocities of the microwave and optical signals. However, some mismatch usually exists, its value can be tailored, and it can be used as an enabling mechanism for EO frequency shifting [59,60].

Let the optical input of an EO phase modulator be a train of pulses that fit into the time slots with duration T and period $2T$ (Figure 5.30). The modulating voltage is a square wave with period $2T$ and magnitude of either V or $-V$, which correspond to two values of the refractive index in the parts of the waveguide where this voltage is applied: $(n_o + \Delta n_o)$ and n_o. The refractive index seen by the microwave signal is $n_m < n_o$. The RF square wave starts to travel in the active region in phase with the optical pulse train. Each optical pulse enters the active region where the index of

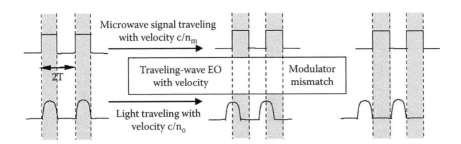

FIGURE 5.30 EO frequency shifting in a polymer phase modulator. Optical and microwave inputs start traveling in the active region in phase, but since microwave propagates faster, at the end they are 180° out of phase. Shaded areas have optical refractive index $n_o + \Delta n_o$ and n_o; areas that are not shaded have refractive index no.

refraction is $(n_o + \Delta n_o)$, but since it travels slower than the microwave signal, it gradually crosses into the region with index n_o. The duration T is chosen so that by the end of the active region the optical pulse train is retarded by exactly T with respect to the modulating square wave, or, equivalently, all parts of the optical signal see index n_o:

$$T = L(n_o - n_m)/c. \tag{5.29}$$

In a Pockels medium, the optical index of refraction n_o depends on the applied electric field E as $n_o(E) \approx n_o - rn_o^3 E/2$, where r is the appropriate Pockels coefficient of the material. In the analysis below, the effects of the electrical loss in the microstrip are excluded for the sake of keeping the calculations tractable. After propagating distance L in a waveguide with voltage V applied to it across electrode spacing d, light undergoes the phase shift of

$$\phi = 2\pi n_o(E)L/\lambda_0 \approx \phi_0 - \pi rn_o^3 EL/\lambda_0 = \phi_0 - \pi V/V_\pi, \tag{5.30}$$

where $\phi_0 = 2\pi n_o L/\lambda_0$, $V_\pi = d\lambda_0 / (Lrn_o^3)$ is the half-wave voltage and λ_0 is the optical wavelength in a vacuum.

Taking into account the velocity mismatch, the time-dependent part of the optical phase at the output of the device can be written as

$$\phi(t) = -\omega \left(t + \frac{1}{n_o - n_m} \frac{rn_o^3}{2d} \int_t^{t+\Delta t} V(\tau)d\tau \right), \tag{5.31}$$

where $\Delta t = (n_o - n_m)L/c$ is the difference between the amounts of time that optical and microwave signals take to cross the active region. The modulating square wave can be expressed as

$$V(t) = \begin{cases} V, & 2kT < t < (2k+1)T \\ -V, & (2k-1)T < t < 2kT \end{cases}, \tag{5.32}$$

where $T = \Delta t$ and k is any integer. At the end of the modulator, the optical pulses occupy time slots for which $(2k-1)T < t < 2kT$. Evaluating the integral in Equation 5.31 for these slots, the final expression for the time-varying part of the optical phase is found to be

$$\phi(t) = -\omega[1 + \Delta n_o/(n_o - n_m)]t. \tag{5.33}$$

Thus, the new optical frequency is

$$\omega_2 = \omega[1 + \Delta n_o /(n_o - n_m)]. \tag{5.34}$$

This confirms that the velocity-mismatched phase modulation of pulsed light with the RF square wave indeed produces EO frequency shifting if the necessary conditions are satisfied.

A signal generator with finite bandwidth cannot create the ideal square wave. In fact, the only RF waveform that can be easily generated with a repetition rate larger than several GHz is a sinusoid. Fortunately, the parts of a sinusoid around its nulls closely approximate a straight line. If optical pulses are aligned to interact with just these parts of a sinusoid, as shown in Figure 5.31, uniform frequency shift can be achieved [61–65].

Substituting $V(t) = V \sin \omega_m t$ into Equation (5.31), performing the integral, and simplifying, we obtain

$$\phi(t) = -\omega \left\{ t + \frac{\Delta n_o L}{2c} \left(\frac{\sin(\omega_m \Delta t/2)}{\omega_m \Delta t/2} \right) \sin(\omega_m t + \omega_m \Delta t/2) \right\}. \tag{5.35}$$

As in the square-wave case, $\Delta n_o = rn_o^3 V/d$ is the peak-to-peak variation in the optical refractive index. The instantaneous frequency can be found by taking the time derivative of phase

$$\omega(t) = \omega \left\{ 1 + \frac{\omega_m \Delta n_o L}{2c} \frac{\sin(\omega_m \Delta t/2)}{\omega_m \Delta t/2} \cos[\omega_m(t + \Delta t/2)] \right\}. \tag{5.36}$$

If the optical pulses are sufficiently short and aligned with the parts of the sinusoid for which $\omega_m(t + \Delta t/2) = \pi q$, where q is an even integer for an upshift and

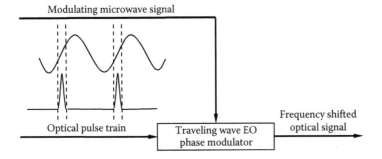

FIGURE 5.31 Conceptual operation of the electro-optic frequency shifter in the upshifting configuration. The microwave frequency has to be an integer multiple of the optical pulse repetition rate.

odd for a downshift, then the last cosine in Equation 5.36 can be approximated as unity for all of the pulse. In this case, there is a uniform frequency shift equal to

$$\Delta\omega = \frac{\omega\,\omega_m\,\Delta n_o\,L}{2c}\,\frac{\sin(\omega_m\Delta t/2)}{\omega_m\Delta t/2}. \tag{5.37}$$

The velocity mismatch in this case actually decreases the frequency shift by the factor of $\mathrm{sinc}(\omega_m\Delta t)$ and thus, unlike the square wave case, is undesirable.

The most natural RF waveform to use for EO frequency shifting is the saw-tooth, since in the absence of velocity mismatch it produces nearly ideal frequency shifts. EO frequency shifting with the saw-tooth is called serrodyne frequency translation [66]. Yet in practice, the shifts achievable by serrodyning are limited to a few MHz, because, as in the square wave case, the finite bandwidth of function generators restricts the highest fundamental frequency of saw-tooth signals they can produce.

It is important to point out that in the presence of velocity mismatch, the actual phase modulation experienced by the optical signal is not equivalent to the shape of the modulating signal. In a velocity-mismatched device, modulation with the properly chosen RF square wave will produce effective phase modulation that is a triangle wave and thus is a linear function of time in the intervals of interest. The equation (5.35) states that for a sinusoidal modulating signal, the effective modulation remains sinusoidal but decreases in amplitude and shifts in phase as the result of the velocity mismatch. Parts of this effective sinusoidal modulation approximate a straight line sufficiently well to be used for EO frequency shifting.

Loss in the microwave electrode is often quite high, so the amplitude of the modulating signal is larger in the beginning of the active region than at the end. This negatively affects the EO frequency shifting. Fortunately, there are ways to cope with this problem. Different microwave waveguide configurations and materials can reduce the losses significantly.

Frequency shifting with a sinusoidal modulating signal has been analyzed and implemented by several groups [61–65]. In [61–63], the oscillator that produced the microwave sinusoid was synchronized with the mode-locker of the laser that generated optical pulses. This approach has three significant disadvantages. First, the optical pulse train and microwave sinusoid have to be produced simultaneously, so the frequency shifter is not transparent to the repetition rate of the optical pulses. Second, the duty cycle of the optical pulses is often quite small, so most of the microwave power does not contribute to the frequency shifting. It does contribute, however, to the heating of the device, thus limiting the maximum power that can be applied. Third, synchronization itself presents a technical challenge, since it often involves generating multiples of high frequencies and is vulnerable to drift and phase noise.

A novel implementation of EO frequency shifting that overcomes these disadvantages was carried out [64,65]. In this approach, the incoming optical pulses generate microwave pulses with a sinusoidal carrier that are later used to modulate the optical pulses. Transparency to the optical pulse repetition rate is obtained, since the microwave and optical pulses are naturally synchronized. High peak powers necessary for large frequency shifts can be achieved, while still keeping the average power well below the device damage threshold.

The diagram of the experimental setup is shown in Figure 5.32. A train of optical pulses from a mode-locked fiber laser with duration of ~200 fs was split into two arms. In the left arm, the pulses were filtered using an arrayed waveguide grating (AWG), amplified in an erbium-doped fiber amplifier (EDFA), passed through a variable delay line (VDL), then into a polarization controller, and finally into a phase modulator. The right arm consisted of a variable optical attenuator (VOA), photodetector (PD), and traveling-wave tube amplifier (TWTA). Thus, in the right arm, each optical pulse was converted into an electrical pulse perfectly synchronized with the optical pulse in the left arm. Group delay in the two arms was equalized using the VDL, so that the optical and electrical pulses entered the modulator simultaneously and interacted in it. After the modulator, the frequency shift was observed on an optical spectrum analyzer. The modulator was fabricated from an EO polymer material consisting of a phenyltetraene bridged high-$\mu\beta$ guest chromophore in an amorphous polycarbonate host [67].

A TWTA with a bandwidth of 8–18 GHz and a peak output power of 10 W was used. Since the TWTA effectively acted as a bandpass filter, the resulting electrical pulse had quasi-sinusoidal oscillations with frequency of ~11 GHz (the central frequency of the TWTA passband weighted for nonuniform gain across the spectrum) and peak-to-peak amplitude of 25 V (Figure 5.33), with the envelope determined by the TWTA bandwidth and dispersion. Thus, the equations developed for sinusoidal modulating signals are relevant in the analysis of this experiment.

The optical filtering with the AWG was necessary because the output spectrum of the mode-locked laser was very broad (~60 nm), much larger than the observed shift.

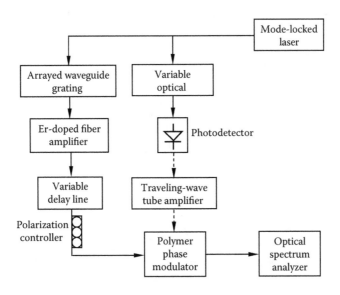

FIGURE 5.32 Block diagram of the experimental setup. The solid lines represent optical signals, and the dashed lines represent electrical signals.

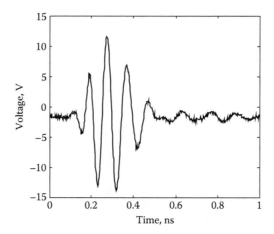

FIGURE 5.33 The temporal shape of the modulating microwave pulse. The dashed ellipse shows the part that downshifts the optical pulse, solid ellipse shows the part that upshifts the optical pulse.

The filtering decreased the optical power in the left arm by 20 dB, which made it desirable to amplify it using the EDFA. Consequently, both AWG and EDFA are not necessary when the frequency conversion of optical pulses with a spectrum typical for optical communications is implemented.

Figure 5.34 shows the initial, upshifted, and downshifted optical spectra. The observed frequency upshift (movement to the left on the wavelength axis) is 86 GHz,

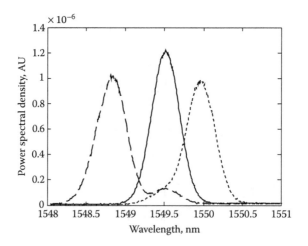

FIGURE 5.34 The result of frequency conversion. Initial spectrum (solid curve), downshifted spectrum (dotted curve), and upshifted spectrum (dashed curve) are shown.

and the frequency downshift is 56 GHz. This correlates well with the simulation predictions when the effects of RF loss are taken into account. The difference between the maximum values of the upshift and the downshift is explained by the difference in the slopes of the sections of the microwave pulse responsible for the upshift and the downshift. The small peak that is evident in the upshifted spectrum is expected and predicted by the simulation, theory, and results [59,61–63]. Frequency shift by any amount smaller than that shown can be easily obtained simply by reducing the input power to the PD using the VOA.

5.9 PHOTONIC RF PHASE SHIFTER FOR OPTICALLY CONTROLLED BEAM-FORMING SYSTEMS

Perhaps the most interesting use of analog polymer devices is in high-frequency phased-array radar systems. Phased-array antenna beam-forming systems have become increasingly important in advanced wireless communications such as mobile radio communications, wireless local area networks (LANs), and satellite communications as well as radars. In contrast to dish or slotted array antennas, which use either physical shape or direction to form and steer beams, phased-array antennas provide highly directional far-field lobe patterns by utilizing the interference of the fields emitted from a large number of regularly spaced antenna elements. The direction of the beams can be steered by electrically controlling the phases of the radio frequency (RF) signals at active phase-shifting elements. However, the requirement of large-scale arrays and high transmission capacity results in the complexity of RF feed structures and a large number of phase-shifting elements. Latest developments in GaAs monolithic microwave integrated circuits (MMIC) technology have helped to reduce these engineering difficulties but still cannot fully satisfy the requirements for practical usage.

Phased-array antennas using photonic radio frequency (RF) phase shifters hold great promise due to their many advantages such as simple implementation, optical distribution capability, low cost, low power consumption, small size, light weight, and immunity to electromagnetic interference [68–73]. Most importantly, in contrast to MMIC systems, the frequency bandwidth of these devices is very wide and effectively covers from DC to over 50 GHz. Also, it allows continuous beam forming without limitation on number of beam angles. Among the possible phase shifter architectures, the one described in [68] is the simplest and most flexible approach. In actual implementation, one can improve the performance with an advanced design and integrate multiple phase shifters in a single chip, providing multiple independent phase outputs. In the following section, the detailed engineering of these structures will be discussed. True time delay systems, which offer extremely wide instantaneous bandwidths, are also being investigated.

Figure 5.35 shows the advanced device architecture for the photonic RF phase shifter. It consists of a carrier-suppressed optical single-sideband (SSB) modulator embedded in one arm of a Mach-Zehnder (MZ) and an optical phase modulator on the other arm. When the SSB modulator unit is driven at quadrature, as shown in Figure 5.35, it typically generates a carrier at Ω and a sideband at $\Omega + \omega_{RF}$ (point A). By inserting an additional balancing arm in the inner MZ, a carrier signal at Ω from the SSB modulator can be suppressed (point B). The desired phase and magnitude of

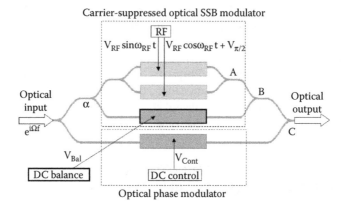

FIGURE 5.35 The basic device architecture for the photonic RF phase shifter.

the optical signal in the balancing arm are established by the proper DC bias, V_{Bal}, and splitting ratio at the balancing arm, α. On the other arm of the MZ, the control DC bias, V_{Cont}, is applied to the optical phase modulator to induce a phase-shifted optical carrier at Ω. Finally, the mixing of these signals in a photodiode gives rise to the RF signal at ω_{RF} with a variable phase controlled by V_{Cont} (point C).

If the input optical signal with unit magnitude at a frequency of Ω is $E_{in}(t) = e^{i\Omega t}$, the expression for the output optical intensity at the modulation frequency ω_{RF} is given by

$$I_{\omega_{RF}}(t) = \frac{1}{4(1+\alpha)^2} A_{RF} J_1(\Delta) \cos(\omega_{RF} t + \varphi_{RF}) \tag{5.38}$$

where

$$\varphi_{RF} = \tan^{-1}\left[\frac{J_0(\Delta) + 2\alpha \sin\phi_{Bal} + 2(1+\alpha)\sin\phi_{Cont}}{J_0(\Delta) + 2\alpha \cos\phi_{Bal} + 2(1+\alpha)\cos\phi_{Cont}}\right], \tag{5.39a}$$

$$A_{RF} = \sqrt{[J_0(\Delta) + 2\alpha \cos\phi_{Bal} + 2(1+\alpha)\cos\phi_{Cont}]^2 + [J_0(\Delta) + 2\alpha \cos\phi_{Bal} + 2(1+\alpha)\cos\phi_{Cont}]^2} . \tag{5.39b}$$

Here V is the half-wave voltage, $\Delta = \pi \cdot V_{RF}/V_\pi$ is the modulation depth, $\phi_{Cont} = \pi \cdot V_{Cont}/V_\pi$ is the optical phase shift by the control DC bias, $\phi_{Bal} = \pi \cdot V_{Bal}/V_\pi$ is the optical phase shift by the balancing DC bias, and α is the optical power-splitting ratio at the balancing arm.

It can be seen from Equation 5.39a that the phase of the detected RF signal can be controlled by changes in control phase, ϕ_{Cont} (or control voltage, V_{Cont}). The balancing phase can be chosen to minimize the power fluctuation in Equation 5.39b, which corresponds to 5/4. This physically indicates that the balancing phase is set to be exactly opposite to the phase of the carrier signal from the SSB modulator unit. In addition, the RF power generated from the photodiode can be determined by Equation 5.38 and turns out to be proportional to A_{RF}^2. The calculated RF phase and power characteristics are shown in Figure 5.36 as a function of control voltage at a modulation depth of 0.5. For the choice of $\alpha = J_0(\Delta)/\sqrt{2}$ with $\phi_{Bal} = 5/4$, the carrier signal from the SSB modulator is fully suppressed and the ideal characteristics for the RF phase and power can be obtained such that

$$\varphi_{RF} = \tan^{-1}\left[\frac{2\sin\phi_{Cont}}{2\cos\phi_{Cont}}\right] = \phi_{Cont}, \qquad A_{RF}^2 = \text{const.}$$

This indicates that the RF power does not vary at all and the RF phase shift is highly linear with respect to the control DC voltage, which makes these devices very suitable for optically controlled phase array antenna systems. Assuming a simple symmetric splitting at the balancing arm, i.e., $\alpha = 1$, with $\phi_{Bal} = 5/4$, this system shows a maximum phase deviation of less than 6° while maintaining the RF power fluctuation below 3 dB as the control voltage is tuned over $2V_\pi$. In this case, the carrier suppression in the SSB modulator is only partially accomplished since the balancing power is slightly unequal to the carrier signal from the SSB modulator unit. Nevertheless, this significant improvement of nearly one order of magnitude is capable of removing one of the major problems in using this type of phase shifter architecture. Note that for the case without the balancing arm ($\alpha = 0$), it starts exhibiting a lack of linearity as the control voltage is tuned over $2V_\pi$. A maximum RF phase deviation of approximately 50° from the ideal linear characteristic is observed. In addition, the RF power exhibits fluctuation of approximately 15 dB as the control voltage is tuned over $2V_\pi$. Most of the detrimental effects are caused by the presence of the strong carrier signal that is not suppressed at all in the SSB modulator unit. This carrier signal is added to another phase-shifted carrier signal at the same frequency Ω and mixed with the sideband in the photodiode. The resulting

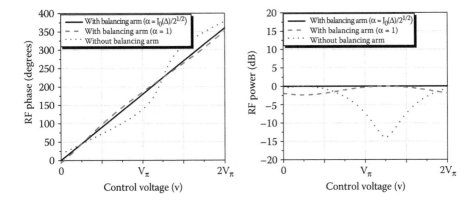

FIGURE 5.36 The calculated RF phase and power characteristics as a function of control voltage at a modulation depth of 0.5.

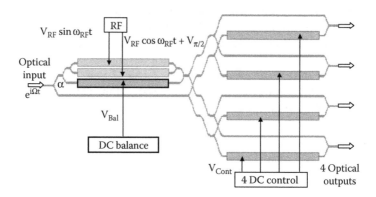

FIGURE 5.37 The schematic diagram for the four-element RF phase shifter array in a single chip.

RF signal reveals degradation of phase and power characteristics. For most practical applications, a wide range of linear phase shifting is required and the RF power fluctuation is very undesirable.

It is favorable to integrate multiple phase shifters in a single chip. This phase shifter array significantly reduces the complexity of RF feed structures and needs only a single RF and optical source. The four-element phase-shifter array incorporating the balanced design is shown in Figure 5.37. The modulated optical output from the balanced SSB modulator is split into four branches and combined with the four outputs from the optical phase shifters. This signal distribution in a planar chip can be achieved through the use of low-crosstalk waveguide crossings. The performance of these devices can be severely impacted by optical waveguide crossings; for this reason, they need to be carefully implemented. Also, the additional propagation loss due to the array structure should be minimized. For this purpose, S-bend shaped waveguides with very low bending losses can be implemented for all bending sections, instead of the regular corner-bend structure, so that the device length can be minimized [74].

Figure 5.38 shows the phase-shifter array in a single chip fabricated using recently developed polymer materials and advanced polymer fabrication technologies, as described above. The device size of the phase shifter with four outputs was

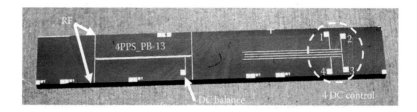

FIGURE 5.38 The balanced multiple-output photonic RF phase shifter fabricated in the APC-CPW polymer material.

3.8 cm × 0.5 cm. For the simplicity of the design, the splitting ratio of the balancing arm, α, was set to be 1.

The measured RF phase and power of a single element are shown in Figure 5.39 at a modulation frequency of 20 GHz and modulation depth of 0.58. Low-frequency triangular waveforms of 50 Hz were used to drive the DC control arms instead of tuning the DC voltages manually. The linear relationship between voltage and time in the control triangular waveform enabled a one-to-one mapping between the measured RF phase (or power) and the control DC voltages. Therefore, the control voltage changes by an amount of $2V_\pi$ for a period of 25 ms. For the control triangular waveforms of $2V_\pi$ $(-12V \sim +12V)$, the RF phase was tuned by 360° with a high level of linearity, and the RF power varied by less than 4 dB, as expected from Figure 5.36. Note that a single control of 360° of the RF phase shift corresponds to the half cycle of the voltage change in triangular waveforms for a period of 25 ms. Accordingly, Figure 5.39 represents 8 times full operation within 200 ms. This performance can be even further improved by employing the design with the optional splitting ratio of the balancing arm as described before.

These RF phase shifters should contain the most important feature of such devices, i.e., that the RF phases of an array element are independently controlled. In order to confirm this, four triangular waveforms of $2V\pi$, set by the equal time delays, were applied to the four DC control arms. The measured RF phase characteristics are shown in Figure 5.40. Almost identical characteristics, including a phase shift of 360°, were observed for all four output ports. It can be also seen that, in a given time frame, this arrangement results in the same effect generated by four different voltages and consequently introduces the independent phase shifts at four output ports. In addition, the independent control of the RF phase was also demonstrated by applying the triangular waveforms having different peak-to-peak voltages to each control arm.

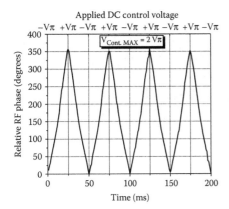
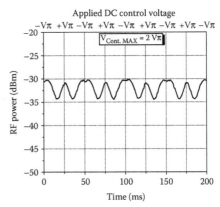

FIGURE 5.39 The measured RF phase and power from a single output.

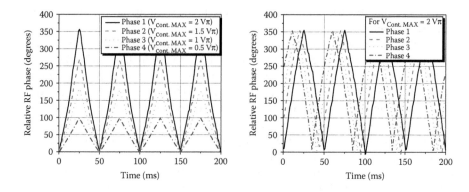

FIGURE 5.40 Independently controlled four-phase outputs.

5.10 RF AMPLIFIER INTEGRATION TO POLYMER DEVICES

In realizing large-scale devices such as the phase shifters presented in the last section, metal conductor losses must be addressed, especially at high frequencies of 40 GHz and beyond. RF amplifier integration to such polymer devices is therefore essential. The first steps towards achieving this are shown by the hybrid integration of a MMIC power amplifier and a polymer phase modulator.

The amplifier used was a three-stage MMIC power amplifier with 23-dB gain at 49.2 GHz and bandwidth 2.5 GHz. Figure 5.41 shows a polymer phase modulator integrated with the RF power amplifier. As shown in Figure 5.42, with 0-dBm RF input power at 49.2 GHz, the integrated device exhibited the first sidelobes 19 dB down from the carrier.

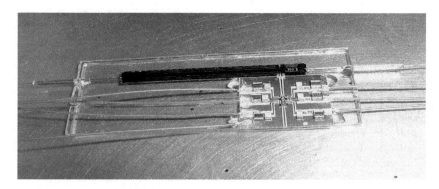

FIGURE 5.41 Photograph of a polymer phase modulator integrated with the RF power amplifier.

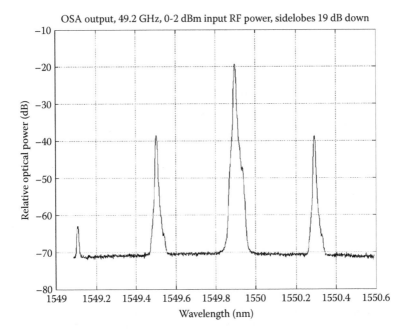

FIGURE 5.42 Optical spectrum of integrated phase modulator device at 49.2 GHz depicting sidelobes 19 dB down from the carrier with 0-dBm RF inputs.

5.11 CONCLUSION

A new generation of high-speed electro-optic polymer devices has been designed and fabricated. They range from ring resonators and flexible devices to DOS switches and directional coupler modulators. Applications include frequency shifters and optically controlled phased-array radars. In each case the use of polymers has unique advantages which make it the technology of choice for the future.

There are, of course, numerous other applications that are currently being explored, including optical digital signal processing and fiber optic gyroscopes. These are potentially vast commercial applications and are stimulating a substantial amount of new development. An example of the first efforts to integrate these polymer devices with drivers was shown in the last section. New devices with a more monolithic approach are now being tested. We look forward to the next stage in the continued progress of high-speed polymer devices.

ACKNOWLEDGMENTS

The authors would like to recognize and thank the many colleagues, Ph.D. students, and research associates who made most of the measurements, fabricated the devices, synthesized the materials, and performed most of the studies described in this chapter. In particular, we would like to acknowledge the work of Kevin Geary from UCLA in organizing, editing, and putting this chapter together. The continued collaboration with the chemistry faculty, in particular Professor Larry R. Dalton, in designing and perfecting the materials was critical. The contributions of all of these people cannot be overemphasized. We would also like to recognize the Air Force Office of Scientific Research, the National Science Foundation, and the Missile Defense Administration for the support that made the work possible. In particular, we want to recognize Dr. Howard Schlossberg and Dr. Charles Y.-C. Lee of AFOSR for their continued support, guidance, and encouragement of this research.

REFERENCES

1. C. C. Teng, "Traveling-wave polymeric optical intensity modulator with more than 40 GHz of 3-dB electrical bandwidth," *Appl. Phys. Lett.*, 60, 1538–1540, 1992.

2. D. Chen, H. R. Fetterman, A. Chen, W. H. Steier, L. R. Dalton, W. Wang, and Y. Shi, "Demonstration of 110 GHz Electro-optic Polymer Modulators," *Appl. Phys. Lett.*, 70, 3335–3337, 1997.

3. M.-C. Oh, H. Zhang, A. Szep, W. H. Steier, C. Zhang, L. R. Dalton, H. Erlig, Y. Chang, B. Szep, and H. R. Fetterman, "Recent advances in electro-optic polymer modulators incorporating phenyltetraene bridged chromophore," *J. Quant Electr. Sel. Top. on Organics for Photonics*, 7, 826–835, 2001.

4. H. Zhang, M.-C. Oh, A. Szep, W. H. Steier, C. Zhang, L. R. Dalton, H. Erlig, Y. Chang, D. H. Chang, and H. R. Fetterman, "Push Pull Electro-optic Polymer Modulators with Low Half-Wave Voltage and Low Loss at both 1310 nm and 1550 nm," *Appl. Phys. Lett.*, 78, 3116–3118, 2001.

5. C. Zhang, A. S. Ren, F. Wang, L. R. Dalton, S.-S. Lee, S. M. Garner, and W. H. Steier, "Thermally Stable Polyene-Based NLO Chromophore and Its Polymers with Very High Electro-Optic Coefficient," *Polymer Preprints,* 40, 49–50, 1999.

6. K. H. Hahn, D. W. Dolfi, R. S. Moshrefzadeh, P. A. Pedersen, and C. V. Francis, "Novel two-arm microwave transmission line for high-speed electro-optic polymer modulators," *Electron. Lett.*, 30, 1220–1222, 1994.

7. Y. Shi, C. Zhang, H. Zhang, J. H. Bechtel, L. R. Dalton, B. H. Robinson, W. H. Steier, "Low (Sub-1-V) halfwave voltage polymeric electro-optic modulators achieved by controlling chromophore shape," *Science*, 288, 119–122, 2000.

8. M. M. Howerton, R. P. Moeller, A. S. Greenblatt, and R. Krahenbuhl, "Fully packaged, broad-band LiNbO$_3$ modulator with low drive voltage," *IEEE Photon. Technol. Lett.*, 12, 792–794, 2000.

9. J. Kondo, A. Kondo, K, Aoki, M. Imaeda, T. Mori, Y. Mizuno, S. Takatsuji, Y. Kozuka, O. S. K. Mohapatra, C. V. Francis, K. Hahn, and D. W. Dolfi, "Microwave loss in nonlinear optical polymers," *J. Appl. Phys.*, 73, 2569–2571, 1993.

10. J. Kondo., A. Kondo, K. Aoki, M. Imadeda, T. Mori, Y. Mizuno, S. Takatsuji, Y. Kozuka, O. Mitomi, M. Minakata, "40-Gb/s X-Cut LiNbO3 Optical Modulator with Two-Step Back-Slot Structure," *J. Lightwave Technol.*, 20, 2110, 2002.

11. Akria Otomo, G. I. Stegeman, W. H. G. Horsthuis, and G. R. Möhlmann, "Strong field, in-plane poling for nonlinear optical devices in highly nonlinear side chain polymers," *Appl. Phys. Lett.*, 65, 2369–2371, 1994.

12. H. Park and W-Y Hwang, "Origin of direct current drift in electro-optic polymer modulator," *Appl. Phys. Lett.*, 70, 2796–2798, 1997.

13. Y. Shi, W. Wang, J. H. Bechtel, A. Chen, S. Garner, S. Kalluri, W. H. Steier, D. Chen, H. R. Fetterman, and L. R. Dalton, "Fabrication and Characterization of High Speed Polyurethane-Disperse Red 19 Integrated Electro-optic Modulators for Analog System Applications," *Sel. Top. Quantum Electr.*, 20, 289–299, 1996.

14. S. Park, J. J. Ju, J. Y. Do, S. K. Park, and M.-H. Lee, "Thermal bias operation in electro-optic polymer modulator," *Appl. Phys. Lett.* 83, 827, 2003.

15. T. C. Kowalczyk, T. Z. Kosc, K. D. Singer, A. J. Beuhler, D. A. Wargowski, P. A. Cahill, C. H. Seager, M. B. Meinhardt, and S. Ermer, "Crosslinked polyimide electro-optic materials," *J. Appl. Phys.*, 5876–5883, 1995.

16. H.-T. Man and H. N. Yoon, "Long term stability of poled side-chain nonlinear optical polymer," *Appl. Phys. Lett.*, 72, 540–542, 1998.

17. W. Sotoyama, S. Tatsuura, and T. Yoshimura, "Electro-optic side-chain polyimide system with large optical nonlinearity and high thermal stability," *Appl. Phys. Lett.*, 64, 2197–2199, 1994.

18. H. Ma, B. Chen, T. Sassa, L. R. Dalton, and K.-Y. Jen, "Highly efficient and thermally stable nonlinear optical dendrimer for electrooptics," *J. Am. Chem. Soc.*, 123, 986–987, 2000.

19. M. A. Mortazavi, H. N. Yoon, and C. C. Teng, "Optical power handling properties of polymeric nonlinear optical waveguides," *J. Appl. Phys.*, 74, 4871–4876, 1993.

20. M. A. Mortazavi, K. Song, H. Yoon, and I. McCulloch, "Optical power handling of nonlinear polymers," *Polymer Preprint*, 35, 198–199, 1994.

21. A. Galvan-Gonzalez, M. Canva, G. I. Stegeman, R. Twieg, T. C. Kowalczyk, and H. S. Lackritz, "Effect of temperature and atmospheric environment on the photo-degradation of some disperse red 1-type polymers," *Optics Lett.*, 24, 1741–1743, 1999.

22. Y. Shi, W. Wang, W. Lin, D. J. Olson, and J. H. Bechtel, "Double-end crosslinked electro-optic polymer modulators with high optical power handling capacity," *Appl. Phys. Lett.*, 70, 1342–1344, 1997.

23. P. Rabiei, W. H. Steier, C. Zhang, and L. R. Dalton, "Polymer Micro-ring Filters and Modulators," *IEEE J. Lightwave Technol.*, 20, 1968–1975, 2002.

24. A. Yariv and P. Yeh, *Optical Waves in Crystals*, Wiley-Interscience, New York, 1984.

25. R. C. Williamson, "Sensitivity-bandwidth product for electro-optic modulators," *Optics Lett.*, 26, 1862–1863, 2001.

26. I.-L. Gheorma and R. M. Osgood, "Fundamental limitations of optical resonator based high-speed EO modulators," *IEEE Phot. Technol. Lett.*, 14, 795–797, 2002.

27. S. T. Kowel, S. Wang, A. Thomsen, W. Chan, T. M. Leslie, and N. P. Wang, *IEEE Photon. Technol. Lett.*, 7, 754, 1995.

28. S. Kalluri, M. Ziari, A. Chen, V. Chuyanov, W. H. Steier, D. Chen, B. Jalali, H. R. Fetterman, and L. R. Dalton, *IEEE Photon. Technol. Lett.*, 8, 644–646, 1996.

29. G. Khanarian, M. A. Mortazavi, and A. J. East, *Appl. Phys. Lett.*, 63, 1462, 1993.

30. L.-M. Wu, Ph. Pretre, R. A. Hill, and A. Knoesen, *Organic Thin Films for Photonic Applications 1997*, OSA Technical Digest Series, 14, 226, 1997.

31. D. Chen, D. Bhattacharya, A. Udupa, B. Tsap, H. R. Fetterman, A. Chen, S. S. Lee, J. Chen, W. H. Steier, and L. R. Dalton, *IEEE Photon. Technol. Lett.*, 11, 54, 1999.

32. Y. Shi, A. Yacoubian, D. J. Olson, W. Lin, and J. H. Bechtel, *Organic Thin Films for Photonic Applications, Optical Society of America, TOPS*, 64, 2002.

33. M.-C. Oh, H. Zhang, C. Zhang, H. Erlig, Y. Chang, B. Tsap, D. Chang, A. Szep, W. H. Steier, H. R. Fetterman, and L. R. Dalton, *IEEE J. Quantum Electron.*, 7, 826, 2001.

34. M. Hikita, S. Tomaru, K. Enbutsu, N. Ooba, R. Yoshimura, M. Usui, T. Yoshida, and S. Imamura, *IEEE J. Quantum Electron.*, 5, 1237, 1999.

35. T. R. Halemane and S. K. Korotky, "Distortion characteristics of optical directional coupler modulators," *IEEE Trans. Microwave Theory Tech.*, 38, 669–673, May 1990.

36. L.M. Johnson and H. V. Roussell, "Reduction of intermodulation distortion in interferometric optical modulators," *Optics Lett.*, 13, 928–930, Oct. 1988.

37. D.J.M. Sabido IX, M. Tabara, T. K. Fong, C.-L. Lu, and L.G. Kazovsky, "Improving the dynamic range of a coherent AM analog optical link using a cascaded linearized modulator," *IEEE Photonics Technol. Lett.*, 7, 813–815, 1995.

38. J.L. Brooks, G. S. Maurer, and R. A. Becker, "Implementation and evaluation of a dual parallel linearization system for AM-SCM video transmission," *J. Lightwave Technol.*, 11, 34–41, 1993.

39. R. B. Childs and V. A. O'Byrne, "Multichannel AM video transmission using a high-power Nd:YAG laser and linearized external modulator," *IEEE Journal on Selected Areas in Communications*, 8, 1369–1376, 1990.

40. R. F. Tavlykaev and R. V. Ramaswamy, "Highly linear Y-fed directional coupler modulator with low intermodulation distortion," *J. Lightwave Technol.*, 17, 282–291, 1999.

41. T. Kishino, R. F. Tavlykaev, and R. V. Ramaswamy, "A Y-fed directional coupler modulator with a highly linear transfer curve," *IEEE Photonics Technol. Lett.*, 12, 1474–1476, 2000.

42. J. H. Schaffner and W. B. Bridges, "Intermodulation distortion in high dynamic range microwave fiber-optic links with linearized modulators," *J. Lightwave Technol.*, 11, 3–6, 1993.

43. R. C. Alferness, "Waveguide electrooptic switch arrays," *IEEE Journal on Selected Areas in Communications,* 6, 1117–1130, 1988.

44. Y. Silberberg, P. Perlmutter, and J. E. Baran, "Digital optical switch," *Applied Physics Lett.*, 51, 1230–1232, 1987.

45. T. A. Birks, P. S. J. Russell, and D. O. Culverhouse, "The acousto-optic effect in single-mode fiber tapers and couplers," *J. Lightwave Technol.*, 14, 2519–2529, 1996.

46. W. K. Burns and A. F. Milton, "Waveguide Transitions and Junctions," *Guided-Wave Optoelectronics*, 2nd ed., Springer-Verlag, 1990.

47. W. K. Burns, "Voltages-length product for modal evolution-type digital switches," *J. Lightwave Technol.*, 8, 990–997, 1990.

48. W. Yuan, S. Kim, D. H. Chang, C. Zhang, W. H. Steier, and H. R. Fetterman, "Electrooptic Polymeric Digital Optical Switches (DOS's) with Low Switching Voltage," *Proc. of CLEO/QELS 2003,* Baltimore, MD, CTuQ5, 2003.

49. R. Krähenbühl, M. M. Howerton, J. Dubinger, and A. S. Greenblatt, "Performance and modeling of advanced Ti:LiNbO3 digital optical switches," *J. Lightwave Technol.*, 20, 92–99, 2002.

50. O. Bryngdhal, "Image formation using self-imaging techniques," *J. Optical Soc. Am.*, 63, 416–419, 1973

51. L. B. Soldano, M. Bachmann, P. A. Besse, and M. K. Smit, "Large optical bandwidth of InGaAsP/ InP MMI 3-dB couplers," presented at the 6th ECIO, Neufchatel, Switzerland, April 1993.

52. M. Bachmann, P. A. Besse, and H. Melchior, "General self-imaging properties in N × N multimode interference couplers including phase relations," *Applied Optics*, 33, 3905–3911, 1994.

53. R.M. Jenkins, J. M. Heaton, D.R. Wight, and J. T. Parker, "Novel 1 × N and N × N integrated optical switches using self-imaging multimode GaAs/AlGaAs waveguides," *Appl. Phys. Lett.,* 64, 684–686, 1994.

54. M. Heaton and R. M. Jenkins, "General matrix theory of self-imaging in multimode interference (MMI) couplers," *IEEE Photonics Tech. Lett.*, 11, 212–214, 1999.

55. L. B. Soldano and E. C. Pennings, "Optical multi-mode interference devices based on self-imaging: Principles and applications," *J. Lightwave Tech.*, 13, no. 4, April 1995.

56. P. A. Besse, M. Bachmann, H. Melchior, L. B. Soldano, and M. K. Smit, "Optical Bandwidth and Fabrication Tolerances of Multimode Interference Couplers," *J. Lightwave Technol.*, 12, 1004–1009, 1994.

57. T. Tamir (Ed.), *Guided Wave Optoelectronics*, Springer-Verlag, 1988.

58. R. Ramaswami and K. N. Sivarajan, *Optical Networks: A Practical Perspective*, Academic Press, San Diego, 1998, pp. 160–166.

59. I. Y. Poberezhskiy and H. R. Fetterman, "Traveling-wave polymer devices as wavelength converters for wavelength-division multiplexing applications," *Optics Lett.*, 27, 2002, 427–429.

60. I. Poberezhskiy, H. R. Fetterman, and D.H. Chang, "Frequency conversion for WDM applications using polymer traveling-wave electro-optic devices," 2001 International Topical Meeting on Microwave Photonics, Technical Digest, pp.105–108.

61. M. A. Duguay and J. W. Hansen, "Optical frequency shifting of a mode-locked laser beam," *IEEE J. Quantum Electronics*, QE-4, 477–481, 1968.

62. M. L. Raziat, G. F. Virshup, and J. N. Eckstein, "Optical wavelength shifting by traveling-wave electrooptic modulation," *IEEE Photonics Technol. Lett.*, 5, 1002–1005, 1993.

63. D. A. Farias and J. N. Eckstein, "Coupled-mode analysis of electrooptic frequency shifter," *IEEE J. Quantum Electron.*, 39, 358–363, 2003.

64. I. Y. Poberezhskiy, B. J. Bortnik, S.-K. Kim, and H. R. Fetterman, "Electrooptic polymer frequency shifter activated by input optical pulses," *Optics Lett.*, 28, 1570–1572, 2003.

65. I. Y. Poberezhskiy, B. J. Bortnik, and H. R. Fetterman, "Quasi-Doppler frequency conversion of optical pulses in traveling-wave polymer phase modulator," CLEO 2003, Baltimore, Md.

66. L. M. Johnson and C. H. Cox, "Serrodyne optical frequency translation with high sideband suppression," *J. Lightwave Technol.*, 6, 109–112, 1988.

67. H. Zhang, M.-C. Oh, A. Szep, W. H. Steier, C. Zhang, L. R. Dalton, H. Erlig, Y. Chang, D. H. Chang, and H. R. Fetterman, "Push-pull electro-optic polymer modulators with low half-wave voltage and low loss at both 1310 and 1550 nm," *Appl. Phys. Lett.*, 78, 3136–3138, 2001.

68. D. K. Paul, "Optical beam-forming and steering for phased-array antenna," *Proc. IEEE Natural Telesys. Conf.* Jun 1993, pp. 7–12.

69. Y. Kamiya, W. Chujo, K. Yasukawa, K. Matsumoto, M. Izutsu, and T. Sueta, "Fiber Optic Array Antenna Using Optical Waveguide Structure," *IEEE Int. Symp. Dig., Antennas and Propagation*, 2072, 774–777, 1990.

70. J. F. Coward, T. K. Yee, C. H. Chalfant, and P. H. Chang, "A photonic integrated-optic RF phase shifter for phased array antenna beam-forming application," *J. Lightwave Technol.*, 11, 2201–2205, Dec 1993.

71. D. Jez, K. Cearns, and P. Jessop, "Optical Waveguide Components For Beam Forming in Phased-Array Antennas," *Microwave and Optical Technol. Lett.*, 15, 46–49, 1997.

72. J. M. Fuster, J. Marti, J. L. Corral, and P. Candelas, "Harmonic up/down-conversion through photonic RF phase shifters in phased-array antenna beam-forming applications," *Microwave and Optical Technol. Lett.*, 22, 247–249, 1999.

73. S. R. Henion and P. A. Schulz, "Electrooptic phased array transmitter," *IEEE Photonics Technol. Lett.*, 10, 424–426, 1998.

74. Jeehoon Han, B. Seo and H. Fetterman, "Integrated polymer photonic RF phase shifters for optically controlled beam forming systems," *Proc. IEEE MTT-S Int. Microwave Symp. (IMS2003)*, June 2003.

6 High-Speed Photodetectors and Photoreceivers

P. Fay

CONTENTS

6.1   INTRODUCTION

High-speed photodetectors and photoreceivers play a critical role in fiber-optic-based telecommunication systems. Conversion of the optical signal in the fiber to an appropriate electronic signal for further processing, while minimizing noise and undesired distortion, is the primary function of photodetectors and receivers in this context. This chapter presents a brief outline of the fundamental physical processes utilized in common photodetectors, followed by a description of some of the most common photodetector types, as well as a discussion of some emerging photodetector designs. In practice, the electrical signal produced by a photodetector alone is often inadequate for practical telecommunication systems; the photodetector must be combined with other optical or electronic amplification and signal processing to form a photoreceiver. The design of high-performance photoreceivers requires careful selection of devices, technologies, and architectures from an increasingly large array of potential choices. The features and limitations of the dominant photoreceiver designs for high-speed fiber-optic communication systems are discussed in the latter portion of the chapter.

6.2 PHOTODETECTORS

Broadly defined, a photodetector is an optoelectronic device that absorbs incident optical energy and converts it to an electrical signal. In general, this can be accomplished in many ways and based on a large number of physical phenomena; the photoelectric effect, intrinsic interband absorption in semiconductors, intersubband absorption in quantum-well heterostructures, and extrinsic absorption mediated by defect and other deep levels are among those that have found wide use. The process of photodetection in a particular device is governed both by the physical processes governing absorption of the incident light and by the transport of electronic charge carriers resulting from the absorption process to produce an external electrical signal. The interplay between optical and electronic material properties, and the degree to which these properties can be engineered to optimize performance, is the basis for recent advances in photodetector technology and permits continued advances in photodetector performance. In the following sections, a brief discussion of the fundamental processes relevant to high-speed photodetectors suitable for telecommunication applications will be presented, along with an introduction to conventional photodetectors, including p-i-n photodiodes, avalanche photodiodes (APDs), and metal-semiconductor-metal (MSM) photodetectors. A discussion of more recent advances in the field, including advanced APDs and resonant-cavity devices, the uni-traveling-carrier photodiode (UTC-PD), and opportunities for future advances will also be presented.

6.2.1 PHOTODETECTOR FIGURES OF MERIT

Regardless of the particular type of photodetector employed, several device figures of merit help to determine the suitability of a given detector for a specific system application. These figures of merit include the detector's responsivity, bandwidth, and noise performance. Responsivity, $\Re$, is a measure of the change in detector current arising from the optical signal; this is conventionally expressed in units of A/W. For many applications it is desirable to maximize the responsivity in order to maximize the overall system's signal-to-noise ratio. Some photodetectors such as avalanche photodiodes exhibit gain, that is, the output photocurrent (in electrons/s) can be larger than the incident absorbed photon flux (in photons/s). This is possible despite the fact that each incident photon produces only one electron-hole pair (EHP), since in an APD the photogenerated carriers undergo impact ionization to generate secondary charge carriers and current multiplication. Although the presence of internal gain can significantly increase the responsivity of a detector, this may be accompanied by a penalty in the bandwidth and noise performance of the detector. Careful consideration of these effects is required to choose the appropriate responsivity-bandwidth-noise tradeoff for a given application.

The operational speed of a photodetector is another important characteristic. This can be specified either in terms of the frequency-domain or the time-domain response. For frequency-domain characterization, the bandwidth is typically specified as the frequency at which the responsivity has fallen off from its low-frequency value by 3 dB. Time-domain approaches characterize the response of the photodetector to very short optical pulses. Often this response is quoted in terms of the full-width at half-maximum (FWHM) of the resulting output electrical pulses. For a detector with a Gaussian impulse response, the bandwidth and FWHM are related by the following expression:

$$f_{3dB} = \frac{2\sqrt{\ln 2}}{\pi \cdot FWHM} \approx \frac{0.533}{FWHM}.$$

In practice, photodetectors do not have purely Gaussian impulse responses, and so caution in this expression's use and interpretation is required. The generally wide disparity in the transport properties of electrons and holes leads to multicomponent impulse responses that often deviate significantly from simple Gaussian pulses. The consequence is that bandwidth estimates based on the assumption of a Gaussian response often significantly overestimate the actual bandwidth of detectors. This overestimation occurs because the FWHM is dominated by the high-velocity carriers, usually electrons. A more complete treatment of the time response by Fourier transformation of the full impulse response (and thus including the "tail" region of the response due to slow carriers) results in bandwidths that match those obtained via frequency-domain techniques.

Another important characteristic of detectors is the amount of noise generated. Noise in detectors can be characterized by the rms noise current, $\langle i^2 \rangle$, the noise equivalent power, NEP, or the specific detectivity, D^*. The noise current spectral density is simply the variance of the photodetector current. When integrated over the desired operating bandwidth, this provides a measure of the rms noise current. The NEP is the required input optical signal power required to make the desired output signal level (rms) equal to the rms noise output in a bandwidth of 1 Hz. The specific detectivity is defined as $D^* = \frac{\sqrt{A \cdot \Delta f}}{NEP}$, where A is the detector area and Δf is the bandwidth. D^* is normally specified in a 1-Hz bandwidth. The choice of normalization for D^* is intended to allow direct comparison of devices of differing areas; however, some care is needed in making comparisons based on D^* since not all detectors follow the $\sqrt{A}$ scaling assumed by this definition of D^*.

Generation of noise is fundamental to the operation of photodetectors and arises from several intrinsic physical phenomena. For most types of photodetectors, the noise generated by the detector can be classified into five main categories — thermal noise, shot noise, generation-recombination noise, flicker or 1/f noise, and excess noise. Thermal noise arises from the random thermal motion of charge carriers inside the device. Shot noise is caused by the discretization of charge composing an electric current as well as by the discrete nature of photons in the optical signal. Fluctuations in the arrival statistics of the individual photons in the optical signal and the electrons that make up the average current give rise to this type of noise. Generation-recombination noise is due to the random timing of EHP generation and recombination events. Thermal noise and shot noise have frequency-independent power spectral densities (a "white" spectral distribution), whereas the spectral density of generation-recombination noise is dependent on the device structure and operational details, but is typically significant only at low frequencies. Flicker noise, often also called 1/f noise after the frequency dependence that it exhibits, is present in detectors as well as all other semiconductor devices. This is generally significant only for frequencies below a few kHz for optimized structures and fabrication processes. Excess noise is any noise contribution of a detector above the limits imposed by the fundamental physical processes of thermal, generation-recombination, and shot noise. This can arise from leakage currents due to defects and unpassivated surface states. In addition, detectors that exhibit gain (e.g., APDs) generally have larger excess noise contributions due to random processes that participate in producing photocurrent gain. Minimization of these excess noise contributions is generally pursued through careful attention to device design, as well as optimization of the fabrication process.

6.2.2 Physical Fundamentals of Photodetectors

Photodetectors for high-speed telecommunications have conventionally been based on intrinsic absorption of photons in semiconductors, generating electron-hole pairs (EHPs). In detectors based on this principle, only photons with energies greater than the semiconductor bandgap are able to generate electron-hole pairs. The electronic response of the detector following electron-hole pair generation is governed by the usual semiconductor charge transport mechanisms of drift and diffusion, and depends strongly on the design and structure of the particular device. For example, in photoconductive detectors the photogenerated carriers give rise to increased conductivity, while in diode-based detectors the photogenerated EHPs give rise to increased terminal current without an appreciable change in conductance.

Photodetectors suitable for high-speed applications are most commonly based on semiconductor p-i-n junctions, although there are a few notable exceptions. Detailed discussion of the operation of photodiodes may be found in a number of excellent texts (see, for example, [1–3]). Fundamentally, photodiodes of this type rely on EHPs generated within the depleted i-region of a reverse-biased p-i-n diode (and also carriers within a diffusion length on either side of the depletion region) for photodetection. Those carriers that are photogenerated within the depletion region are swept through the high-field depletion region by the electric field due to the applied bias and built-in potential, while the carriers that are generated in the p- and n-type regions must first diffuse to the depletion region before being swept across the depletion region. These moving charges give rise to a transient displacement current at the diode terminals, as well as contributing a drift or conduction current.

These drift and diffusion processes govern the intrinsic speed of the detector. For example, if photogenerated carriers can be swept across the depletion region more quickly, or if the diffusive component can be reduced or eliminated, intrinsic device speed can be improved. In addition to these intrinsic limitations, the speed of a detector is limited by extrinsic factors as well. Extrinsic effects include, for example, the time needed to charge and discharge device junction capacitances through the external load resistance (or the input impedance of a following circuit or amplifier). In practice, both intrinsic and extrinsic factors must be considered, since the intrinsic and extrinsic mechanisms are often comparable in time scale. In many cases these factors depend on the same physical or geometric factors, albeit with differing functional dependence. In addition to the physical processes that produce photodetection, thermal and field-assisted generation of carriers in the depletion region contribute to the device current, independent of illumination, which contributes to photodetector noise. Noise in detectors will be treated in more detail in subsequent sections.

6.2.3 MATERIAL SYSTEMS

The preceding qualitative discussion is generally applicable to junction photodiodes in any semiconductor material system. The material system of choice for a given application depends on system parameters such as the wavelength(s) to be detected, the detection bandwidth required, and noise considerations. The ability to attain required system-level performance targets is closely related to fundamental optical and electronics material properties. For high-performance photodetectors, there is often a need to incorporate materials that are transparent at the detection wavelength, as well as materials to function as the absorption medium. For example, a common approach to eliminating slow diffusive effects in p-i-n photodiodes is to use wide bandgap, optically transparent p- and n-regions with a narrow bandgap i-region as the absorption layer. Thus photogeneration can only take place in the high-field undoped i-region, and diffusive transport from the contact layers is eliminated. Transparent materials are also required for the realization of antireflective coatings on detector surfaces, for core and cladding layers in integrated guided-wave structures, and as passivation layers to prevent environmental processes from degrading photodetector performance over time.

The electronic properties of materials are also important for realizing high-speed photodetectors. These are of course intimately related to the optical properties since they too arise fundamentally from the material bandstructure. For photodetector design, the primary electronic properties of interest include the conduction and valence band dispersion relations (usually through conduction and density-of-states effective masses and band nonparabolicity), the low-field electron and hole mobilities, and the high-field transport properties. In addition, the density and distribution of trapping states in the material can impact detection performance as well. All of these properties originate from the details of the atomic arrangement of the material's constituent atoms and so have an inherent dependence on mechanical properties as well. Strain affects both the optical and electronic properties, and the amount of strain that can be accommodated without relaxation in a structure is limited. Thus, candidate materials must be selected for epitaxial compatibility as well, primarily by ensuring that the materials are either in lattice match or are thin enough to be coherently strained.

For telecommunications applications, the transmission properties of optical fiber dictate the choice of optical wavelengths. Conventional single-mode fiber has two low-loss "windows" in the vicinity of 1.3 μm and 1.55 μm, separated by a loss peak due to water within the fiber. More recently, fiber with greatly reduced water content has been introduced, with a continuous low-loss window from 1.34 μm to 1.63 μm [4]. Since the cutoff wavelength for optical absorption is governed by $\lambda_c = \frac{hc}{E_g}$, where h is Planck's constant, c is the speed of light in vacuum, and E_g is the material bandgap, materials with bandgaps in the vicinity 0.75 eV are required. For high-speed, long-haul telecommunications, this limits consideration to InGaAs, Ge, and other materials with comparatively small bandgaps. Of course, for other applications for which minimizing fiber loss is not such a dominant factor (e.g., short-haul fiber links, free-space optical interconnects, etc.) or for extremely cost-sensitive applications where performance is secondary to cost, other materials (e.g., GaAs, Si) may be preferable since in these cases other wavelengths (e.g., 850 nm) are often selected. The bandgap, in conjunction with other band structure parameters, also contributes to the determination of the intrinsic carrier concentration, n_i, and to the thermal generation rate of EHPs within the material. This strongly influences the conduction and noise properties of photodetectors; a small bandgap generally leads to a larger thermal generation rate and increased band-to-band tunneling in high-field regions. These effects lead to increased dark current and thus to increased detector noise.

The details of the conduction and valence band dispersion relations also play a significant role in controlling the behavior of photodetectors. Fundamentally the generation of EHPs by photon absorption, and thus the absorption coefficient α, is controlled by the availability of filled valence band states and unoccupied conduction band states separated by the incident photon energy under the constraint of conservation of momentum. Thus, the density of states in both the conduction and valence band, as well as the details of these states' distributions in energy and in momentum, must be considered. Due to the extremely small photon momentum in the 1.3- to 1.55-μm wavelength range, the conservation of momentum favors absorption in direct-gap materials. Materials such as Si and Ge, which have indirect bandgaps, require additional phonon interactions to provide the necessary momentum for photon absorption at wavelengths near the band edge. This more complicated process results in a generally lower absorption coefficient, and thus it requires a larger absorption volume for comparable efficiency to a direct-gap material. The larger size can adversely affect device speed for high-speed applications. Higher-energy photons may also be absorbed in indirect bandgap materials through transitions between the valence band and higher direct-gap conduction bands.

While the bandgap places effective upper limits on the wavelengths that can be absorbed in a given material, the combination of surface states and increasing absorption coefficient with decreasing wavelength tend to limit the lower wavelength limit [5]. The absorption coefficient increases with increasing photon energy due to the increased density of states at energies away from the conduction and valence band edges. Thus, light incident on a semiconductor surface is absorbed over a shorter penetration distance into the material. Due to dangling bonds and other imperfections on typical semiconductor surfaces, EHPs generated near the surface may recombine before being collected at the electrodes, reducing detector quantum efficiency.

The combination of increased recombination at surfaces and decreased absorption distance at shorter wavelengths, as well as the increased dark current mentioned previously, tends to favor the use of absorbing materials with bandgaps that are only modestly smaller than the incident photon energy for best efficiency and noise. It should be noted that detector designs specifically engineered to ensure drift-dominated transport away from the illuminated surface have been demonstrated to substantially improve short-wavelength performance [5].

This combination of factors favors the use of InGaAs/InAlAs/InP for high-speed detectors in long-haul fiber-based systems since InGaAs provides a good match to the required bandgap (0.74 eV), while InAlAs and InP provide lattice-matched materials with larger bandgaps for controlling the absorption in the structure as well as good electron transport properties.

6.3 PHOTODETECTOR TYPES

6.3.1 P-I-N PHOTODIODES

Variations of the p-i-n photodiode are perhaps the most common type of high-speed photodetector. Although p-i-n photodiodes can in principle be operated in the photovoltaic mode [6], charge storage effects result in low bandwidths when used in this way. For high-speed applications, they are operated with an external reverse-bias voltage applied sufficient to fully deplete the intrinsic region. The resulting photocurrent is then used as the output signal. While p-n junctions can be used as photodetectors, devices of this type suffer from poor quantum efficiency due to the small absorption volume defined by the width of the depletion region. In addition, these detectors generally have modest bandwidths because they have a comparatively large junction capacitance, again due to the small depletion width. The p-i-n photodiode can be viewed as an enhancement of the p-n junction that alleviates these two difficulties through the addition of an undoped or lightly doped intrinsic (i) region between the p- and n-type regions, resulting in a p-i-n structure. The doping level of the intrinsic region is chosen low enough so that for relatively small reverse-bias voltages the undoped region is fully depleted. As a consequence, significant enlargement of the absorption volume is possible with structures of this kind, leading to improved quantum efficiency. Furthermore, the junction capacitance is decreased, leading to larger detection bandwidth. An approximate analysis of the p-i-n homo-junction photodiode structure illustrated in Figure 6.1(a) yields the following expression for responsivity (after [1]):

$$\Re = \frac{q\lambda}{hc}(1-R)\eta_i\left(1-\frac{e^{-\alpha d}}{1+\alpha L_p}\right)$$

where d is the thickness of the depletion region, η_i is the internal quantum efficiency, R is the surface reflectivity, α is the absorption coefficient, and L_p is the diffusion length for minority carrier holes. This expression is derived assuming that the applied

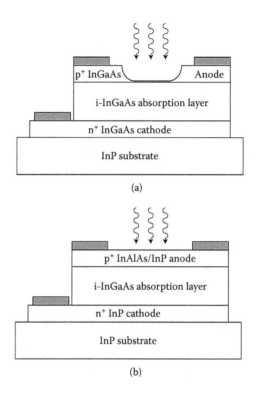

FIGURE 6.1 Schematic cross-sectional diagram of (a) homojunction p-i-n photodiode and (b) heterojunction p-i-n diode with wide-bandgap anode and cathode layers.

bias voltage is sufficient to fully deplete the intrinsic region and yet is sufficiently small so as to avoid breakdown effects, and that negligible absorption takes place in the p-region. This latter condition can be satisfied by etching a window in the p-region as shown in Figure 6.1(a), or by ensuring that the thickness of the p-region in the optical path is much less than $1/\alpha$. Use of a heterojunction p-i-n structure as shown in Figure 6.1(b) eliminates absorption in the wide bandgap p- and n-regions entirely, and thus eliminates the αL_p term in the denominator of this expression. This eliminates the need to etch a window or otherwise restrict the p-layer thickness.

Similarly, an idealized analysis of the high-frequency response of these devices yields an approximate expression for detector bandwidth:

$$f_{3dB} = \frac{1}{2\pi\sqrt{(R_L C_j)^2 + \tau_t^2}}$$

where C_j is the device junction capacitance, R_L is the external load resistance (or input resistance of the next circuit stage), and τ_t is the photocarrier transit time. As can be seen in this expression, junction capacitance and carrier transit time primarily control detector bandwidth. Junction capacitance in mesa-structure devices such as those shown in Figure 6.1 is usually well approximated by the parallel-plate capacitance expression $C_j = \varepsilon A/d$. The capacitance can be reduced by using a thicker absorption layer or by reducing diode cross-sectional area. The transit time of photogenerated carriers across the depletion region τ_t has a more complex dependence on photodetector structure, as it depends not only on depletion layer thickness, but also on electron and hole velocities and the spatial distribution of EHP generation. In simple top-illuminated designs, the EHP generation rate decays as $e^{-\alpha d}$ from the top of the illuminated InGaAs surface. Additionally, for semiconductors of practical interest, electron and hole velocities at a given electric field strength are often significantly different. The spatial inhomogeneity of EHP generation and asymmetric transport complicates the expression for τ_t somewhat (see, for example, [2]). The fundamental trade-off between transit time and charging time is unchanged by these complications, however. Both the transit time and the junction capacitance are controlled by the thickness of the intrinsic region; increases in i-layer thickness result in decreased capacitance but increased transit time. For an optimal design, the junction capacitance charging and transit-time contributions to the bandwidth are set approximately equal, and thus a unique solution for optimum intrinsic region thickness for a given detector area from bandwidth considerations can be found. Since the responsivity also increases with increasing i-layer thickness, this highlights the contradictory requirements of high responsivity and bandwidth in the p-i-n photodiode structure.

The noise performance of p-i-n photodiodes is limited by generation-recombination noise, dark current shot noise, and resistive thermal noise. In high-speed devices, 1/f and generation-recombination noise can often be neglected since they predominate at low frequencies. Neglecting these effects, the rms noise current is given approximately by

$$\langle i^2 \rangle = \int_0^{\Delta f} \left[\frac{4k_B T}{R} + 2q\,(I_D + I_p) \right] df$$

where the total rms noise current from the detector is found by integrating the noise current spectral densities from shot and thermal noise over the electrical bandwidth, Δf, of interest. It should be pointed out that these expressions for responsivity, bandwidth, and noise relate to the small-signal linear, analog application of the detectors. Under large-signal conditions due to high illumination levels, photogenerated carriers may screen the external applied electric field, significantly complicating the frequency response. A full treatment of these effects can be found in [7]. Extension of these results to the prediction of bit error rates in digital systems also requires some further refinements, primarily to account for the nonstationary nature of shot noise in a digital system (e.g., [8]).

6.3.2 ADVANCED P-I-N PHOTODIODES

The inherent trade-off between responsivity and bandwidth in conventional p-i-n photodiodes has driven consideration of a wide variety of potential alternatives.

The fundamental limitation that these more advanced structures seek to overcome is that in normally incident p-i-n photodiodes, increases in responsivity require thicker intrinsic absorption regions, but this is diametrically opposed to increasing bandwidth due to the increase in transit time and subsequent decrease in bandwidth. A sampling of advanced p-i-n structures is shown schematically in Figure 6.2(a)–(d). Figure 6.2(a) depicts a resonant-cavity structure; in this structure, distributed Bragg reflectors (DBR) are grown above and below the absorption region. In comparison to the conventional p-i-n structures shown in Figure 6.1 in which the incident light passes through the absorbing layer only once, these Bragg reflectors act as mirrors, bouncing the unabsorbed light back and forth in the absorption region and increasing the effective length of the intrinsic region for the optical field. At the same time, the transit length for photogenerated EHPs is typically reduced since a thin absorbing region at the center of the resonant cavity is used [9–10]. In practice, the additional access resistance imposed by the DBR heterobarriers can limit somewhat the achievable improvement in bandwidth. This resonant structure also produces spectral selectivity due to the Fabry-Perot modes of the resonant cavity structures.

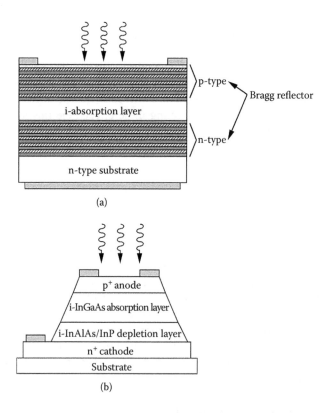

FIGURE 6.2 Schematic illustrations of p-i-n photodiode types. (a) Resonant-cavity p-i-n photodiode. (b) Dual-depletion p-i-n photodetector. (c) Waveguide photodiode. (d) Velocity-matched traveling-wave photodiode.

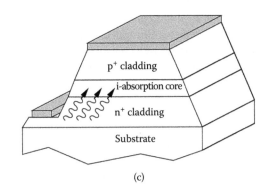

(c)

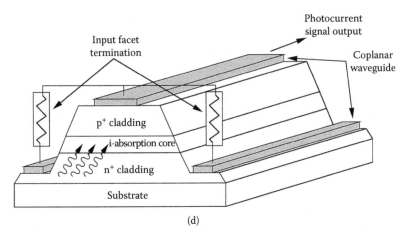

(d)

FIGURE 6.2 (Continued)

Instead of attempting to improve the responsivity, another approach to improve p-i-n photodiode performance is the use of the dual-depletion layer structure [11]. A representative schematic of this approach is shown in Figure 6.2(b). Fundamentally, this approach seeks to improve detector bandwidth by decoupling the RC charging time from the photocarrier transit time. This design makes use of the fact that in most III-V materials of interest, the electron transport is much faster than hole transport. In this case, introducing a modest increase in the distance that electrons must travel does not significantly alter the transit time (which is limited by holes). However, it can significantly lower the device capacitance, since the depletion layer consists in this case of both the absorption layer and the second depletion layer on the cathode side. This second depletion layer must be optically transparent to prevent holes from being photogenerated in that region and thus increasing the hole transit time. This requires the use of a heterojunction in the depletion region, which may cause deleterious charge pile-up effects at the interface [12]. These effects can be significantly reduced through the use of graded layers (see, for example, [13]).

Another approach to improve the bandwidth-responsivity product is the edge-illuminated waveguide photodetector illustrated in Figure 6.2(c). In this device, a guided-wave structure with an absorptive core is used, and the light is input at the edge of the detector (see, for example, [14–16]). Since the incident light propagates horizontally along the waveguide and the photocarriers travel vertically between the top and bottom electrodes, the optical and electrical path lengths are decoupled. As a consequence, the light can be absorbed over a longer optical path length than with the conventional normally incident design. Coupling losses must be carefully controlled, however, since the acceptance angle and cross-sectional area of the photodetector waveguide are often small. This limits the achievable external quantum efficiency, although the internal quantum efficiencies of devices of this type can be nearly unity. Since in this device the absorption path length and the intrinsic region thickness are decoupled, a short transit time can be maintained while still increasing the absorption length. One limitation of this structure, however, is that the length of the waveguide can be quite long for high quantum efficiency. Although the transit time of the EHPs is kept short, the long device length leads to a large junction capacitance, resulting in an RC-limited bandwidth.

Figure 6.2(d) shows a simplified version of a traveling-wave photodetector that can overcome this difficulty. In this structure, the top electrode is formed into an electrical coplanar waveguide (CPW), whose propagation velocity is matched to the optical propagation velocity in the integrated optical waveguide. In this way, a modulated input optical signal and the electrical signal that it generates through EHP generation travel away from the input facet at the same velocity. This circumvents the effects of the excess capacitance in the conventional waveguide detector by distributing it along a CPW, and it has been shown to result in extremely wide detection bandwidths [17,18]. The CPW termination near the input facet is used to reduce the occurrence of reflections from the facet end of the CPW. This termination has the undesirable effect of reducing the responsivity (one-half of the photocurrent from EHP generation is dissipated in the termination, with the remaining one-half available at the output end of the CPW) as well as contributing additional thermal noise to the output signal, but it significantly reduces pulse distortion in the detector. As with the conventional waveguide detector, input optical coupling losses must be carefully controlled in order to achieve high external quantum efficiency. A conceptually similar device is the evanescently coupled velocity-matched distributed photodetector [19,20]. In devices of this type, multiple discrete photodiodes are placed at intervals along a passive optical waveguide. A phased-matched electrical waveguide interconnects the photodiodes. The evanescent coupling allows high optical powers to be detected without exciting detector nonlinearity by distributing the optical power over a large number of detectors.

6.3.3 Avalanche Photodiodes

Avalanche photodiodes (APDs) are another class of photodetector that finds wide use in telecommunication systems. Based on the p-i-n photodiode structure, APDs offer advantages for high-sensitivity communication systems by providing internal photocurrent gain via avalanche carrier multiplication. This improvement in responsivity is not without cost, however, since APDs are more complicated to manufacture than conventional p-i-n photodiodes, require high bias voltages (typically 25 to 40 V) that must be carefully regulated and temperature compensated, and may suffer from both speed and noise penalties associated with the dynamics of the avalanche gain process if not designed carefully. Recent advances in materials growth and manufacturing as well as commoditization of bias control circuits have significantly eased these barriers, so that APDs are no longer solely used for long-haul networks, but are emerging in cost-sensitive metro-area fiber networks as well.

The design of APDs is a particularly rich and active area of research, with many possible design options and potential material choices. The simplest implementation of an APD is a p-i-n photodiode operated under sufficiently high reverse bias so as to cause the photogenerated EHPs to impact ionize in the depletion region. Since the choice of materials that can be used for the absorption region is constrained by the wavelength to be detected, compromise in the gain performance to achieve adequate absorption performance is inevitable in this simple structure. The noise performance of such a simple device in most relevant compound semiconductors is poor. Consequently, structures that separate absorption and gain processes are generally favored. It can be shown that the best avalanche multiplication noise performance is achieved in materials for which avalanche processes are dominated by one carrier type only [1,3]. In order to afford the device designer the freedom to choose both the optimal absorption material and a suitable one-carrier-dominant avalanche region material, device structures of the form shown in Figure 6.3(a) can be used. In this structure, the separate absorption/multiplication APD (SAM-APD), the gain and absorption processes occur in separate and distinct layers in the device. As a practical matter, it is important to manage the band discontinuities between the absorption layer and the multiplication layer to avoid the deleterious effects of charge pile-up at this heterointerface.

One difficulty with the SAM-APD structure shown in Figure 6.3(a) is that the electric field in the multiplication region and the absorption region are tightly coupled. Since impact ionization is desired only in the larger-bandgap multiplication region, but typically varies as $E_g^{-3/2}$ [2], this suggests that it would be preferable to decouple the electric field in the absorption region from that in the multiplication region. A modest field sufficient for high drift velocity but insufficient for significant impact ionization is preferable for the absorption region, while a high field to rapidly accelerate carriers to initiate ionization is desired in the multiplication region. The separate absorption-charge-multiplication APD (SACM-APD), illustrated in Figure 6.3(b), is an effective method for implementing this basic strategy.

Multiplication regions based on InAlAs, InP, and even Si (through wafer bonding) have been used with $In_{0.53}Ga_{0.47}$ As absorption layers for 1.3–1.55 μm wavelength communication systems [21–26], as well as other material systems for other wavelengths.

The added complexity associated with designing and controlling APDs is made up for in high-sensitivity communications systems by their intrinsic current gain.

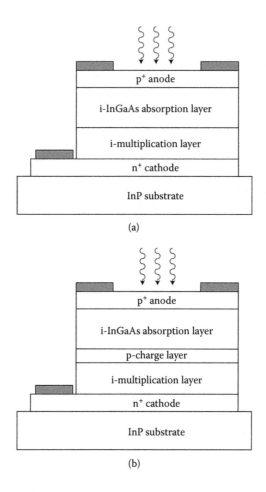

FIGURE 6.3 Schematic cross-sectional diagram of avalanche photodiodes designed for electron multiplication. (a) Separate absorption and multiplication APD (SAM-APD), and (b) separate absorption-charge-multiplication structure (SACM-APD) to permit different electric fields in the absorption and multiplication layers. The structures shown are for electrons as the favored carrier type for avalanche gain. Structures designed for hole multiplication place the multiplication layer on the anode side of the absorption layer, and use an n-type charge layer instead.

Since typical p-i-n-based photoreceivers are electronic amplifier noise limited (see Section 6.4.1), the current gain in an APD can significantly improve the sensitivity by providing a larger input signal to the preamplifier. However, since impact ionization is fundamentally a random process, excess noise is generated along with the current gain due to random fluctuations in the timing of the impact ionization events as well as the number of impact ionization events, and thus the gain itself is randomized. The noise performance of APDs is also a function of the gain; the classical excess noise factor, F, for SAM-APDs can be expressed as [1]

$$F_n = \frac{\beta_p}{\alpha_n}G_n + \left(1 - \frac{\beta_p}{\alpha_n}\right)\left(2 - \frac{1}{G_n}\right)$$

$$F_p = \frac{\alpha_n}{\beta_p}G_p + \left(1 - \frac{\alpha_n}{\beta_p}\right)\left(2 - \frac{1}{G_p}\right)$$

where F_n is the excess noise for the case when electrons are injected into the multiplication region and F_p corresponds to the injection of holes. G_n and G_p are the avalanche gain for each of these two situations, α_n is the impact ionization coefficient for electrons, and β_p is the coefficient for holes. As can be seen in these expressions, the excess noise factor is reduced for materials with large disparities between the impact ionization coefficients for electrons and holes. The noise current spectral density for an APD, including the effects of multiplication noise, is given approximately by

$$\langle i^2(f)\rangle = 2q(I_D + I_p)GF + \frac{4k_BT}{R}$$

where G is the multiplication gain and R is the shunt resistance of the APD. The total noise current is found by integrating this noise current spectral density over the electrical bandwidth of interest.

The excess noise factors above arise from traditional assumptions and simplifications, primarily the assumption of a purely local dependence between the electric field and the impact ionization probability. For devices intended for use at bit rates of 2.5 Gbit/s and below, relatively thick absorption and multiplication layers can be used without significant carrier transit time effects. The push to higher bandwidths and bit rates, however, forces a reduction in multiplication layer thickness to submicron dimensions. It has long been recognized that these traditional assumptions are not valid for very thin layers, and it has recently been demonstrated that the use of thin multiplication layers can produce much lower excess noise factors than the classical analysis suggests [27].

The origin and full explanation of the improved noise performance for thin high-speed APDs is an active area of research. Several physical mechanisms appear to contribute to the observed low noise level in these devices. Two effects that have been suggested are (1) "dead zone" effects and (2) hot-carrier injection into the multiplication layers. The dead zone arises from the fact that carriers must accelerate to achieve sufficient energy to initiate impact ionization. While for thick multiplication layers the distances over which this acceleration takes place is negligible, the thinner layers needed for high-speed devices can become comparable to this dead-zone thickness. Although the dead zone reduces the avalanche gain, it also reduces excess noise by lowering the uncertainty in the position in the multiplication region at which impact ionization can take place, thereby reducing the photocurrent variance. To fully account for this effect, the history of the carriers must be considered, and thus this requires a nonlocal model for the ionization process [28]. In addition to the dead zone, it is possible in heterojunction devices to inject carriers into the multiplication region with nonnegligible kinetic energy due to band bending at the heterointerface. This hot-carrier injection may reduce the thickness of the initial dead space prior to the first impact ionization event of a photocarrier, resulting in a further reduction in noise as well as an increase in gain due to more efficient use of the multiplication layer thickness. A more detailed treatment of the noise and bandwidth effects associated with thin-layer APDs can be found in [28–30].

The internal, potentially low-noise, gain of APDs is very attractive for high-speed telecommunication systems to improve receiver sensitivity. Since the APD gain precedes the electronic preamplifier, this gain can usefully improve sensitivity for electronic-noise limited systems. At bit rates of 2.5 and 10 Gbit/s, the APD is well-established for long-haul telecommunications, and it is increasingly being deployed in shorter-reach links. An active area of research as of this writing is whether the gain-bandwidth product of these devices can be expanded to allow significant gains with adequate bandwidth for 40-Gbit/s class signals. To date, a record gain-bandwidth product of 320 GHz has been obtained with an RC limited bandwidth of 28 GHz [31], which shows promise for this technology.

6.3.4 ADVANCED APD DESIGNS

As with p-i-n photodiodes, numerous advanced structures have been developed to improve device performance, in addition to the refinements represented by the SAM-APD and SACM-APD discussed previously. As in p-i-n photodiodes, APDs have been implemented in resonant cavity structures to enhance responsivity and provide spectral selectivity [32,33] as well as in edge-illuminated waveguide configurations [31].

Unique to the APD is an additional opportunity for optimizing performance based on the idea of engineering a multiplication layer rather than using the properties of an available bulk semiconductor. As described in the previous section, the excess multiplication noise in an APD can be minimized if the impact ionization is dominated by only one carrier type. For most compound semiconductors, however, $k = \beta_p/\alpha_n \sim 1$, and even for the InP multiplication layers used in commercially viable devices $k = 0.4$ [34]. To further improve k and thus improve the noise performance, artificial or composite materials specifically engineered for high asymmetry in impact ionization are attractive. One approach to realizing highly asymmetric impact ionization is to grade the multiplication layer to favor ionization events for one carrier type over the other. This graded multiplication layer APD approach, while difficult in practice due to the usual difficulties in growing well-controlled graded epitaxial layers, has been demonstrated to be capable of producing $k < 0.1$ in a GaAs-based design for detection at 850 nm [35]. This approach and other engineered or structured materials approaches appear particularly promising for future advances in APDs.

6.3.5 Metal-Semiconductor-Metal Photodetectors

An alternative to p-i-n based detectors is the metal-semiconductor-metal photodetector (MSM-PD). In this type of detector, planar Schottky rectifying contacts are made in an interdigitated geometry to the absorption layer. A schematic illustration for the InAlAs/InGaAs/InP material system is shown in Figure 6.4. For InGaAs-based MSM-PDs, the extremely low Schottky barrier height obtained for most metals on InGaAs necessitates the inclusion of an additional barrier enhancement layer, typically of InAlAs, to reduce the dark current [36]. Since the photogenerated carriers must traverse this barrier enhancement layer to be collected by the electrodes, mitigation of heterobarrier interface effects are especially important in MSM-PDs. Careful control of the InGaAs/InAlAs interface, typically through compositional grading of a quaternary InAlGaAs layer, is required to avoid introducing a potential barrier in the photocurrent path. The heterobarrier can also be reduced by short-period superlattice grading of the Schottky enhancement layer from the low-bandgap absorption layer to the larger-bandgap Schottky layer. Counter-doping the Schottky layer has also been shown to be effective at minimizing carrier pile-up effects [37]. Since both electrodes are rectifying contacts, under bias the device consists of one forward-biased Schottky diode and one reverse-biased diode. In the absence of photogenerated EHPs, the saturation current of the reverse-biased contact governs the device current. Under illumination, the photogenerated EHPs in the absorption region drift under the influence of the fringing electric field established by the externally applied bias to produce a photocurrent.

The interdigitated electrode geometry is almost universally utilized for MSM-PDs, since it permits very low device capacitances to be achieved. In contrast to the

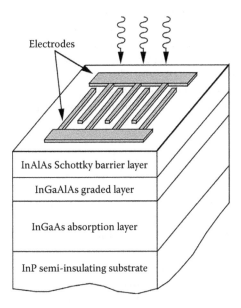

Electrodes

InAlAs Schottky barrier layer

InGaAlAs graded layer

InGaAs absorption layer

InP semi-insulating substrate

FIGURE 6.4 Structure of a typical metal-semiconductor-metal photodetector.

approximately parallel-plate geometry of both p-i-n and APD based detectors, the fringing-dominated electric field distribution associated with an interdigitated electrode geometry exhibits a much lower capacitance. This provides the possibility of a larger RC-limited bandwidth. In addition, this geometry allows high electric fields to be established with modest bias voltages, while simultaneously permitting straightforward control of the transit length of photogenerated carriers. It should be noted that while p-i-n based detectors have vertical current transport, MSM-PDs are primarily a lateral transport device. Thus, the carrier transit length is controlled through lithography, while the quantum efficiency is controlled independently through epitaxial layer design. In addition to the low capacitance, an additional advantage of the MSM-PD structure is its relative ease of fabrication.

For the geometry shown in Figure 6.4, the detector is generally illuminated from the top surface. In this configuration, the metal electrodes block a fraction of the light from entering the semiconductor and thus reduce the maximum achievable external quantum efficiency by a factor associated with the "fill factor"

$$\Gamma = \frac{w}{s+w}$$

of the electrodes, where s is the spacing between adjacent electrodes and w is the width of the electrodes. Although the interdigitated electrode configuration makes MSM-PDs appear similar to some types of photoconductors, the rectifying contacts in MSM-PDs preclude the reinjection of unrecombined EHPs. This eliminates photoconductive gain and limits the internal quantum efficiency of the MSM-PDs to unity or less. In practice, however, the high electric fields present along the semiconductor surface in the interdigitated geometry can give rise to surface-state mediated gain mechanisms. Gain has also been shown to arise from image force lowering effects due to charge accumulation at potential barriers [38]. The dynamics of such defect-assisted mechanisms are generally quite slow, limiting the usefulness of this mechanism for improving responsivity. Neglecting these gain effects and assuming top-side illumination and opaque electrodes, the photocurrent in an MSM-PD under sufficient bias to fully deplete the absorption region is given approximately by the expression

$$\Re = \frac{q\lambda}{hc}(1-R)\eta_i(1-e^{-\alpha d})(1-\Gamma).$$

The external quantum efficiency can in principle be improved by through-wafer illumination to avoid the fill-factor quantum efficiency penalty. However, the non-uniform electric field in the MSM-PD, and particularly the very low electric field present directly beneath the electrodes, contributes to a dramatic reduction in transit-time-limited bandwidth for through-wafer-illuminated MSM-PDs. In the case of through-wafer illumination, carriers generated beneath the electrodes must first diffuse to a higher-field region or an electrode before collection. This serial process of diffusion and drift limits the use of through-wafer-illuminated MSM-PDs to applications with modest bandwidth requirements.

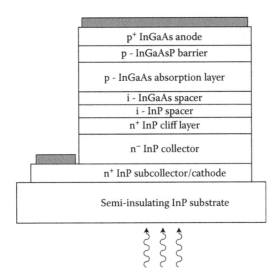

FIGURE 6.5 Simplified layer structure for a uni-traveling-carrier photodiode (UTC-PD).

6.3.6 Uni-Traveling-Carrier Photodiode

Another device with notable performance for high-speed systems is the uni-traveling-carrier photodiode (UTC-PD) [39]. A representative device structure of one implementation of this device is shown in Figure 6.5. The operation of this device departs in a number of significant ways from conventional photodiodes, and has demonstrated extremely large bandwidth of >150 GHz [40] and very large output signal swings of >1.5 V at 40 Gbit/s [41]. In contrast to more conventional structures, the UTC-PD uses a p-type doped absorption layer, followed by an InGaAsP diffusion barrier layer. Under illumination, the EHP generation in the p-absorption layer dramatically increases the electron concentration, but does not significantly alter the hole concentration due to the p-doping. Electrons are extracted from the absorption region and accelerated into the InP cliff layer, using carefully selected doping levels to engineer the electric field for maximum drift velocity. In this way, only electrons significantly contribute to the photocurrent, and thus slow hole transit times are avoided. The typically thin absorption layer thickness limits the detector's responsivity in a normally-incident geometry, but edge-illuminated configurations have achieved improved responsivity (0.48 A/W) [41]. The extremely high speeds and large output signal swings make this an interesting device and an attractive candidate for many systems.

6.4 PHOTORECEIVERS

For many applications, the output photocurrent signal available directly from a photodetector is inadequate to meet system requirements. Additional amplification, conversion to a signal voltage at a different typically lower impedance level, or other processing such as filtering and pulse shaping is required to convert the signal to a more usable form. In a long-haul telecommunications application, very low-level optical signals may be present at the input to the photoreceiver, requiring that the signal be significantly amplified prior to logic thresholding and conversion to an explicitly digital electronic signal. In short-haul links, the use of a photoreceiver can significantly reduce the launched optical power required, and thus lower cost. The ability to work with low-level optical signals also reduces the deleterious effects of nonlinear fiber propagation that can be caused by launching large optical powers into fiber spans. A more complete introduction to these issues may be found in [4].

Although the basic functionality of a photoreceiver is to provide some level of amplification and filtering or pulse shaping, more advanced system functions are sometimes incorporated into the photoreceiver. These functions include logic thresholding, clock recovery, and retiming. Whether these functions are considered part of the photoreceiver or are considered to be part of a later system block varies with one's perspective. The focus of this chapter will be on the universal amplification and filtering properties of photoreceivers, since these impose fundamental limits on link performance.

For telecommunications applications, the ultimate goal is to achieve an acceptably low bit error rate (BER) at the desired bit rate. The performance of the photoreceiver plays an important role in determining the BER of the system, and a photoreceiver's impact on system BER is characterized by several device figures of merit. These figures of merit include the photoreceiver gain or responsivity (typically in V/W), the group delay flatness, dynamic range, bandwidth, and ultimately the sensitivity. Of course, the individual devices or components (transistors, photodetectors, etc.) that make up the photoreceiver are also characterized by their gain, bandwidth, noise figure, and other metrics; it is the interplay between these individual component performance metrics that governs the photoreceiver sensitivity and ultimately contributes to determining the system BER.

Independent of the detailed selection of devices and circuit topologies for photoreceiver implementation, some basic trade-offs and rules of thumb are useful in framing a discussion of photoreceivers. For example, the electrical bandwidth of the photodetector and photoreceiver is a critical parameter for determining the sensitivity of the photoreceiver. As the bandwidth is increased, the noise level at the photoreceiver's output also increases due to the larger bandwidth for thermal and shot noise. As the bandwidth is decreased, the output noise decreases, but also the likelihood of bit errors due to intersymbol interference (ISI) is increased. In practice, a bandwidth of at least 0.8 times the bit rate is required to avoid excessive ISI-related BER penalties for non-return-to-zero (NRZ) coding. Thus, noise considerations and ISI considerations place opposing pressures on photoreceiver bandwidth, and the bandwidth must be optimized to provide the best overall photoreceiver performance [8]. Of course, in the detailed design of a photoreceiver, the gain, noise, and bandwidth are a complex function of the particular devices selected, the impedance environment each device experiences, and the circuit topology or photoreceiver architecture.

In addition to effects due to device performance, the performance of a particular photoreceiver is also controlled by external system-level design choices. For example, the way in which the optical signals are encoded has a significant effect on the sensitivity and performance of a photoreceiver. Traditionally, on-off keying (OOK) has been used in fiber-optic systems due to its simplicity. In this coding strategy, the light intensity is modulated between on and off to signal a 1 or 0. OOK is implemented in two main flavors — return-to-zero (RZ) and non-return-to-zero (NRZ). In the NRZ approach, the light signal is set to the on or off state for the entire duration of the bit period. In the RZ scheme, a short light pulse within a bit period denotes a 1 and the absence of a pulse denotes a 0. In this scheme, the optical signal always returns to 0 between each bit, since the pulses are significantly shorter than the bit period. The choice of RZ or NRZ has significant ramifications on photoreceiver and system performance. Due to the narrower pulses inherent to RZ coding, photoreceivers for RZ systems must have a larger electrical bandwidth to pass the on-state pulses without excessive attenuation. On the other hand, RZ coding is more robust against common fiber impairments [42,43] and is less susceptible to intersymbol interference due to limited receiver bandwidth effects. A more complete introduction to data-coding schemes and their impact on link performance can be found in [8].

Although simple, OOK-based approaches suffer from fundamental limitations of optical spectral efficiency. Historically, spectral efficiency has not been a significant concern for fiber-optics based communications, but with the continued growth in data traffic and the desire to continue decreasing cost-per-bit in telecommunication systems, spectral efficiency is beginning to become important and is poised to take on increasing significance in the future. For wavelength-division multiplexed (WDM) systems, optical spectral efficiency translates directly into the required optical channel spacings needed to avoid excessive bit errors due to crosstalk between channels. More advanced coding schemes, including duobinary signaling [44–46], differential phase-shift keying (DPSK) [47], and differential quadrature phase-shift keying (DQPSK) [48] offer substantial advantages, at the cost of being more complex to implement. These schemes typically require more complex receivers; for example, one way to implement a DPSK receiver is shown in Figure 6.6. This receiver uses a balanced receiver architecture preceded by a delay interferometer. The benefits, however, can be significant — the sensitivity of such a receiver is 3 dB better than a comparable single-ended receiver [47]. To date, more complex spectrally-efficient schemes such as varieties of QAM that are routinely employed at RF have been

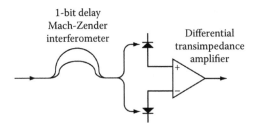

FIGURE 6.6 Balanced receiver for DPSK-modulated data. The 1-bit delay Mach-Zender interferometer converts the phase-modulated data to amplitude modulation for detection by the photodiodes.

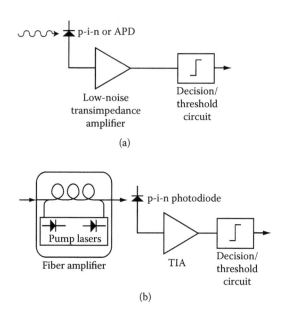

FIGURE 6.7 Illustrations of (a) p-i-n and APD-based electronically amplified photoreceivers and (b) an optically amplified (using a fiber amplifier in this case) photoreceiver.

used only in modest per-channel bandwidth systems such as video distribution systems [49]; the bit rates of fiber-based telecommunication systems are much higher than typical wireless channels, and thus the circuitry required for these coding schemes is difficult to implement.

As one might expect, there are many ways to implement the functionality required of a photoreceiver. Figure 6.7 shows a selection of the possibilities, including electronic amplification and optical preamplification. In addition, hybrid designs are also possible and may offer advantages. For example, even an optically amplified receiver with optical gain sufficiently high to make electronic noise insignificant may benefit from additional electronic gain or amplitude control in order to provide convenient signal and impedance levels to subsequent circuit blocks. Additionally, the use of an electronic preamplifier after the photodetector can improve bandwidth by providing a low input impedance to the detector (thus reducing the RC time constant of the photodetector) and presenting a low output impedance to the subsequent electronic stages. The trade-offs and limitations of each of these approaches are somewhat different; each is treated in the subsequent sections.

6.4.1 Electronically Amplified Receivers

Photoreceivers based on electronic amplification can be implemented in many ways. Typical configurations include p-i-n photodiodes in conjunction with high-speed heterojunction bipolar transistors (HBTs) [50–52] or high electron mobility transistors (HEMTs) [53–56]. Other configurations that have been demonstrated include

MSM-PDs with HEMTs [57,58] and APDs with a number of electronic device technologies [59,60]. For typical telecommunication applications, high sensitivity at the desired bit rate is a key design goal. To achieve this, a photodetector with high external quantum efficiency is needed. A high-quantum-efficiency photodetector provides a large photocurrent signal to the electronic preamplifier, thus improving the overall photoreceiver sensitivity.

An additional consideration in the design of a high-speed electronics-based photoreceiver is the minimization of ISI. This requires the use of a photodetector with adequate transit-time-limited bandwidth, the selection of preamplifier input impedance low enough to keep the detector RC time constant short compared to the bit period, and design of the electronic preamplifier for sufficient gain and group-delay flatness. This last requirement often limits the use of inductive peaking in the amplifier since such peaking techniques can produce large 3-dB bandwidths but impose large start-up and decay times for transient signals such as are present in digital data streams. The photodetector and amplifier must also be either DC coupled or have a very low (typically just a few kHz) corner frequency to avoid causing bit errors due to long strings of 0s for RZ-coded data and both 1s and 0s for NRZ data. Finally, the electronic preamplifier must be designed for low circuit noise. Sensitivity of electronically-amplified photoreceivers is typically dominated by electronic circuit noise, so any improvements made in preamplifier noise level benefit the overall photoreceiver performance. This requirement, in conjunction with the bandwidth requirements for the desired bit rate, drives the selection of device technologies to those offering good low-noise performance, for example, low-noise III-V HEMTs and either SiGe or III-V HBTs.

Additional options for photoreceiver design arise from the variety of fabrication techniques that are available to integrate the components of the photoreceiver. Hybrid assembly, in which a discrete photodetector is electrically connected to a high-speed preamplifier and other circuitry, continues to be a viable method of making photoreceivers, even as the bit rates have continued to increase. Much research has also been devoted to developing monolithically integrated photoreceivers, with the photodetector and electronic amplifier fabricated on a common substrate. This monolithic approach has potential advantages, particularly at high bit rates, since it largely eliminates interconnect parasitics between the photodetector and the electronic preamplifier. These parasitics can reduce bandwidth and introduce distortion, and monolithic integration offers a path to significantly reduce these problems. On the other hand, hybrid techniques offer the ability to choose a photodetector and amplifier independent of their material system or device technologies, and thus offer the ability to choose "best of breed" devices for each function in the receiver. This flexibility comes at the cost that these devices are discrete components, which must be packaged and interconnected to each other without incurring unacceptable performance penalties. This requires significant effort in packaging to preserve signal integrity at high bit rates. To date, monolithic and hybrid receivers have demonstrated nearly equal sensitivity performance, provided the device technologies used to implement the receivers is the same. The difficulty in implementing APDs monolithically with transistor technologies, however, has restricted the use of these detectors to hybrid photoreceiver applications. Due to the internal gain of the APD, hybrid

APD-based receivers can provide superior sensitivities to p-i-n-based monolithic photoreceivers, demonstrating the advantage obtained by the added flexibility of hybrid approaches.

Photoreceiver sensitivity is limited by several noise processes, including photon and electronic shot noise, and electronic device noise. Analyses of the fundamental physical limits of sensitivity have reported various limits, including 10 photons/bit [8] to 36–38 photons/bit [61,62]. These analyses assume a BER of 10^{-9} with NRZ OOK signaling. The limits vary primarily due to what physical effects each author considers to be "fundamental" noise sources. For a 1.55-μm wavelength, the conservative 38-photon/bit limit corresponds to a sensitivity of −43 dBm at 10 Gbit/s and −37 dBm at 40 Gbit/s. In practice, even the best electronically amplified receivers are not able to approach particularly close to this fundamental limit. Instead, the dominant noise process in this type of receiver is the noise contributed by the electronic pre-amplifier. For p-i-n based photoreceivers, sensitivities approximately 20 dB above this limit are more typical [50,51,53]. In order to improve the sensitivity of electronically-amplified receivers, reductions in circuit noise or techniques to increase the signal level at the input of the amplifier are required. This emphasizes the need for photodetectors with high external quantum efficiencies and makes detectors with internal gain mechanisms such as APDs particularly attractive. This also explains the limited appeal of MSM-based receivers, since the finite fill factor on these devices depresses the external quantum efficiency below that of comparable p-i-n based designs [57].

6.4.2 APD-Based Receivers

As alluded to in the preceding section, photodetectors with gain are especially attractive variants on the electronically amplified photoreceiver. Since the gain in an APD is essentially an electronic process rooted in impact ionization, receivers based on APDs are governed by the same fundamental limitations as other electronically amplified photoreceivers. However, since the gain of the APD precedes the electronic preamplifier circuit, the APD supplies a larger input signal to the transistor-based preamplifier. Consequently, the electronic preamplifier noise figure is less significant in determining the photoreceiver sensitivity, and improved noise performance can result. The actual degree to which the sensitivity is improved depends on the performance of the APD — the gain of the APD is not noiseless, but has associated noise due to the random nature of avalanche multiplication, as discussed in Section 6.3.3. For state-of-the-art APDs with bandwidths suitable for 2.5 Gbit/s and 10 Gbit/s applications, the excess noise associated with the avalanche gain is sufficiently low as to provide significant improvements in photoreceiver sensitivity with the inclusion of an APD, compared to a p-i-n photodiode of comparable quantum efficiency, with APD-based receiver sensitivities of −29.5 dBm having been reported at 10 Gbit/s [60]. This sensitivity is improved by approximately the APD current gain, compared to typical p-i-n-based receivers.

Despite these improvements in sensitivity over other electronics-limited photoreceivers, it should be noted that the sensitivity is still approximately 13 dB away from the fundamental sensitivity noise limit. The gain of APDs with sufficient bandwidth for 10 Gbit/s telecommunication applications is typically on the order of 10, while in order to achieve the fundamental noise-limited performance an additional 1 to 2 orders of magnitude is required for typical electronics noise levels [8]. Active research to further improve the gain-bandwidth product appears promising to improve the performance achievable with APD-based receivers, as well as to open up higher bit rate systems (e.g., 40 Gbit/s) to APD receiver options.

6.4.3 Optically Amplified Receivers

A natural extension of the principle of including low-noise gain as early in a receiver system as possible drives the consideration of moving gain out of the electronic domain entirely, and into the optical domain prior to photodetection. Such optically amplified photoreceivers have been enabled by the semiconductor optical amplifier (SOA) as well as by fiber-based amplifiers such as erbium-doped fiber amplifiers (EDFAs). Regardless of the type of optical amplifier used, optically amplified receivers present a larger optical signal to the photodetector and subsequent electronic circuitry. This larger optical signal lessens the significance of the noise contributions made by any electronic amplification or signal processing stages. Further, the optical gain reduces, and in some cases may eliminate entirely, the need for electronic amplification (e.g., [63]). In systems of this type, the large optical signals made possible by optical amplification permit the photodetector to be directly connected to electronic decision/logic threshold circuits, without intervening gain stages.

As with other gain mechanisms, however, optical amplifiers do not provide noiseless gain. The primary source of noise in both SOAs and EDFAs is amplified spontaneous emission (ASE) noise. An excellent review of noise in optical amplifiers can be found in [62]. ASE ultimately limits the noise figure of these amplifiers to a lower bound of 3 dB [8,62]. Since ASE-related noise is broadband, achieving the best sensitivity in an optically amplified photoreceiver requires an optical channel filter between the amplifier and photodetector to remove out-of-band ASE noise. However, the high gains obtainable with optical amplifiers in many cases more than makes up for the ASE noise and any filter losses. Optical gain of more than 30 dB is not uncommon for fiber-based optical amplifiers [62], and with gains of this size the contribution to sensitivity of the electronic circuitry and photodetector noise becomes much less significant [8,62]. This significantly improves the sensitivity of photoreceivers based on this technology, and it allows photoreceiver sensitivity to approach the fundamental quantum limits on BER performance. For example, at 10 Gbit/s a sensitivity within 2.5 dB of the quantum limit for NRZ and 1.4 dB for RZ coded data has been achieved using a low-noise EDFA coupled to a p-i-n-based photoreceiver [64].

Although the sensitivity performance of optically amplified photoreceivers has been demonstrated to be excellent, for some applications there are some drawbacks. Fiber-based amplifiers are comparatively large in size, and require relatively high-power pump lasers to achieve high gains. By contrast, SOAs are much smaller and have even been shown to be monolithically integrable with photodetectors [65], resulting in a much more compact photoreceiver module. However, SOA noise figures are typically larger than those of optimized fiber-based amplifiers, and the monolithic integration of ASE filters is challenging [65]. This has limited the sensitivity achievable with SOA-based receivers. Advances in integration technology and SOA performance promise to further improve the performance of receivers of this type.

6.5 CONCLUSIONS

Photodetector and photoreceiver design is a dynamic field, with an active worldwide research community continuing to advance the state of the art. Continued advances in photodetector designs, including improvements in high-speed APDs and advanced structures like the UTC-PD and others, are particularly exciting and promise substantial benefits to fiber-based communication systems as well as other applications. Similarly, advances in photoreceiver design, including the inclusion of optical gain within the photoreceiver module, have been shown to be effective in improving performance. Further advancements in integration and component performance are promising for realizing high-performance, compact, low-power, and low-cost receivers for ever-wider deployment of high-capacity telecommunication systems.

REFERENCES

1. S. L. Chuang, *Physics of Optoelectronic Devices*, John Wiley & Sons, New York, 1995.
2. P. Bhattacharya, *Semiconductor Optoelectronic Devices,* 2nd. ed., Prentice Hall, Upper Saddle River, NJ, 1997.
3. W. T. Tsang, ed, "Lightwave Communications Technology, Part D" in *Semiconductors and Semimetals*, Vol. 22, Academic Press, Orlando, FL 1985.
4. G. Keiser, "Optical Fiber Communications," in *Wiley Encyclopedia of Telecommunications*, John Wiley & Sons, New York, 2003.
5. Y. Sun, J. C. Campbell, S. Wang, A. L. Beck, A. Chen, A. Yulius, and J. M. Woodall, "Drift-Dominated InP Photodetectors with High Quantum Efficiency," *Proc. Intl. Conf. on Indium Phosphide and Related Materials*, 2003, pp. 502–505.
6. A. G. Dentai, C. R. Giles, E. Burrows, C. A. Burrus, L. Stulz, J. Centanni, J. Hoffman, and B. Moyer, "A Long-Wavelength 10-V Optical-to-Electrical InGaAs Photogenerator," *IEEE Photonics Technol. Lett.*, 11, 114–116, 1999.
7. K. J. Williams, R. D. Esman, and M. Dagenais, "Nonlinearities in p-i-n Microwave Photodetectors," *IEEE J. Lightwave Technol.*, 14, 84–96, 1996.
8. P. J. Winzer, "Optical Transmitters, Receivers, and Noise," in *Wiley Encyclopedia of Telecommunications*, John Wiley & Sons, New York, 2003.
9. G. Kinsey, C. Lenox, H. Nie, J. C. Campbell, and B. G. Streetman, "Resonant Cavity Photodetector with Integrated Spectral Notch Filter," *IEEE Photonics Technol. Lett.*, 10, 1142–1144, 1998.
10. T. Knodl, H. K. H. Croy, J. L. Pan, R. King, R. Jager, G. Lullo, J. F. Ahadian, R. J. Ram, C. G. Fonstad Jr., and K. J. Ebeling, "RCE Photodetectors Based on VCSEL Structures," *IEEE Photonics Technol. Lett.*, 11, 1289–1291, 1999.
11. F. J. Effenberger and A. M. Joshi, "Ultrafast, Dual-Depletion Region, InGaAs/InP p-i-n Detector," *IEEE J. Lightwave Technol.*, 14, 1859–1864, 1996.

12. X. Li, N. Li, X. Zheng, S. Demiguel, J. C. Campbell, D. A. Tulchinsky, and K. J. Williams, "High-Saturation-Current InP-InGaAs Photodiode with Partially-Depleted Absorber," *IEEE Photonics Technol. Lett.*, 15, 1276–1278, 2003.

13. J. H. Jang, G. Cueva, W. E. Hoke, P. J. Lemonias, P. Fay, and I. Adesida, "Metamorphic Graded Bandgap InGaAs/InGaAlAs/InAlAs Double Heterojunction P-i-I-N Photodiodes," *IEEE J. Lightwave Technol.*, 20, 507–514, 2002.

14. K. Kato, A. Kozen, Y. Muramoto, Y. Itaya, T. Nagatsuma, and M. Yaita, "110-GHz, 50%-Efficiency Mushroom-Mesa Waveguide p-i-n photodiode for a 1.55 μm Wavelength," *IEEE Photonics Technol. Lett.*, 6, 719–721, 1994.

15. A. Umbach, D. Trommer, G. G. Mekonnen, W. Ebert, and G. Unterborsch, "Waveguide Integrated 1.55 μm Photodetector with 45 GHz Bandwidth," *Electronics Lett.*, 32, 2143–2145, 1996.

16. M. Shishikura, H. Nakamura, S. Tanaka, Y. Matsuoka, T. Ono, T. Miyazaki, and S. Tsuji, "High-Responsivity, Low-Dark-Current InGaAlAs Waveguide Photodiode with a Symmetric Double-Core for Optical Access Networks," *Electronics Lett.*, 32, 1882–1883, 1996.

17. K. S. Giboney, R. L. Nagarajan, T. E. Reynolds, S. T. Allen, R. P. Mirin, M. J. W. Rodwell, and J. E. Bowers, "Travelling-Wave Photodetectors with 172-GHz Bandwidth and 76-GHz Bandwidth-Efficiency Product," *IEEE Photonics Technol. Lett.*, 7, 412–414, 1995.

18. K. S. Giboney, M. J. W. Rodwell, and J. E. Bowers, "Traveling-Wave Photodetector Design and Measurements," *IEEE J. Selected Topics Quantum Electronics*, 2, 622–629, 1996.

19. L. Y. Lin, M. C. Wu, T. Itoh, T. A Vay, R. E. Muller, D. L. Sivco, and A. Y. Cho, "Velocity-Matched Distributed Photodetectors with High-Saturation Power and Large Bandwidth," *IEEE Photonics Technol. Lett.*, 8, 1376–1378, 1996.

20. M. S. Islam, S. Murthy, T. Itoh, M. C. Wu, D. Novak, R. B. Waterhouse, D. L. Sivco, and A. Y. Cho, "Velocity-Matched Distributed Photodetecdtors and Balanced Photodetecdtors with p-i-n Photodiodes," *IEEE Trans. Microwave Theory and Techniques*, 49, 1914–1920, 2001.

21. J. Yu, L. E. Tarof, R. Bruce, D. G. Knight, K. Visvanatha, and T. Baird, "Noise Performance of Separate Absorption, Grading, Charge, and Multiplication InP/InGaAs Avalanche Photodiodes," *IEEE Photonics Technol. Lett.*, 6, 632–634, 1994.

22. I. Watanabe, M. Tsuji, K. Makita, and K. Taguchi, "Gain-Bandwidth Product Analysis of InAlGaAs-InAlAs Superlattice Avalanche Photodiodes," *IEEE Photonics Technol. Lett.*, 8, 269–271, 1996.

23. G. S. Kinsey, C. C. Hansing, A. L. Holmes Jr., B. G. Streetman, J. C. Campbell, and A. G. Dentai, "Waveguide In0.53Ga0.47As-In0.52Al0.48As Avalanche Photodiode," *IEEE Photonics Technol. Lett.*, 12, 416–418, 2000.

24. B. F. Levine, A. R. Hawkins, S. Hiu, B. J. Tseng, C. A. King, L. A. Gruezke, R. W. Johnson, D. R. Zolnowski, and J. E. Bowers, "20 GHz High Performance Planar Si/InGaAs p-i-n Photodetector," *Proc. Intl. Conf. on Indium Phosphide and Related Materials*, 1997, pp. 483–485.

25. A. Pauchard, M. Bitter, D. Sengupta, Z. Pan, S. Hummel, Y. H. Lo, Y. Kang, P. Mages, and K. L. Yu, "High-Performance InGaAs-on-Silicon Avalanche Photodiodes," *Proc. Optical Fiber Communication Conf.*, 2002, pp. 345–346.

26. J. Wei, C. Dries, H. Wang, M. L. Lange, G. H. Olsen, and S. R. Forrest, "Optimization of 10-Gbit/s Long-Wavelength Floating Guard Ring InGaAs-InP Avalanche Photodiodes," *IEEE Photonics Technol. Lett.*, 14, 977–979, 2002.

27. M. A. Saleh, M. M. Hayat, P. P. Sotirelis, A. L. Holmes, J. C. Campbell, B. E. A. Saleh, and M. C. Teich, "Impact-Ionization and Noise Characteristics of Thin III-V Avalanche Photodiodes," *IEEE Trans. on Electron Devices*, 48, 2722–2731, 2001.

28. M. M. Hayat, O.- H. Kwon, S. Wang, J. C. Campbell, B. E. A. Saleh, and M. C. Teich, "Boundary Effects on Multiplication Noise in Thin Heterostructure Avalanche Photodiodes: Theory and Experiment," *IEEE Trans. on Electron Devices*, 49, 2114–2123, 2002.

29. P. J. Hambleton, B. K. Ng, S. A. Plimmer, J. P. R. David, and G. J. Rees, "The Effects of Nonlocal Impact Ionization on the Speed of Avalanche Photodiodes," *IEEE Trans. on Electron Devices*, 50, 347–351, 2003.

30. M. M. Hayat, O.- H. Kwon, Y. Pan, P. Sotirelis, J. C. Campbell, B. E. A. Saleh, and M. C. Teich, "Gain-Bandwidth Characteristics of Thin Avalanche Photodiodes," *IEEE Trans. on Electron Devices*, 49, 770–781, 2002.

31. G. S. Kinsey, J. C. Campbell, and A. G. Dentai, "Waveguide Avalanche Photodiode Operating at 1.55 μm with a Gain-Bandwidth Product of 320 GHz," *IEEE Photonics Technol. Lett.*, 13, 842–844, 2001.

32. C. Lenox, H. Nie, P. Yuan, G. Kinsey, A. L. Holmes, B. G. Streetman, and J. C. Campbell, "Resonant-Cavity InGaAs-InAlAs Avalanche Photodiodes with Gain-Bandwidth Product of 290 GHz," *IEEE Photonics Technol. Lett.*, 11, 1162–1164, 1999.

33. H. Nie, K. A. Anselm, C. Lenox, P. Yuan, C. Hu, G. Kinsey, B. G. Streetman, and J. C. Campbell, "Resonant-Cavity Separate Absorption Charge and Multiplication Avalanche Photodiodes with High-Speed and High Gain-Bandwidth Product," *IEEE Photonics Technol. Lett.*, 10, 409–411, 1998.

34. J. C. Dries, "Indium Gallium Arsenide Avalanche Photodiodes...Not Just for Telecom Anymore," *Proc. 2002 IEEE GaAs Symp.*, 2002, pp. 51–54.

35. S. Wang, R. Sidhu, X. G. Zheng, X. Li, X. Sun, A. L. Holmes, and J. C. Campbell, "Low-Noise Avalanche Photodiodes with Graded-Impact-Ionization-Engineered Multiplication Region," *IEEE Photonics Technol. Lett.*, 13, 1346–1348, 2001.

36. H. T. Griem, S. Ray, J. L. Freeman, and D. L. West, "Long-Wavelength (1.0–1.6 μm) In0.52Al0.48As/In0.53(GaxAl1-x)0.47As/In0.53Ga0.47As Metal-Semiconductor-Metal Photodetector," *Appl. Phys. Lett.*, 56, 1067–1068, 1990.

37. J.-I. Chyi, Y.-J. Chien, R.- H. Yuang, J.- L. Shieh, J.- W. Pan, and J.- S. Chen, "Reduction of Hole Transit Time in GaAs MSM Photodetectors by p-type δ-Doping," *IEEE Photonics Technol. Lett.*, 8, 1525–1527, 1996.

38. J. Burm and L. F. Eastman, "Low-Frequency Gain in MSM Photodiodes Due to Charge Accumulation and Image Force Lowering," *IEEE Photonics Technol. Lett.*, 8, 113–115, 1996.

39. T. Ishibashi, S. Kodama, N. Shimizu, and T. Furuta, "High-Speed Response of Uni-Traveling Carrier Photodiodes," *Jpn. J. Appl. Phys.*, 36, part 1, no. 10, pp. 6263–6268, 1997.

40. N. Shimizu, N. Watanabe, T. Furuta, and T. Ishibashi, "InP-InGaAs Uni-Traveling-Carrier Photodiode with Improved 3-dB Bandwidth of Over 150 GHz," *IEEE Photonics Technol. Lett.*, 10, 412–414, 1998.

41. H. Fukano, Y. Muramoto, K. Takahata, and Y. Matsuoka, "High Efficiency Edge-Illuminated Uni-Travelling Carrier Structure Refracting-Facet Photodiodes," *Electronics Lett.*, 35, 1664–1665, 1999.

42. L. Boivin and G. J. Pendock, "Receiver Sensitivity for Optically Amplified RZ Signals with Arbitrary Duty Cycle," *Proc. Optical Amplifiers and their Applications*, 1999, pp. 106–109.

43. P. J. Winzer and A. Kalmar, "Sensitivity Enhancement of Optical Receivers by Impulsive Coding," *IEEE J. Lightwave Technol.*, 17, 171–177, 1999.

44. A. Lender, "The Duobinary Technique for High-Speed Data Transmission," *IEEE Trans. on Communication Electronics*, 82, 214–218, 1963.

45. S. Aisawa, J. Kani, M. Fukui, T. Sakamoto, M. Jinno, S. Norimatsu, M. Yamada, H. Ono, and K. Oguchi, "A 1580-nm Band WDM Transmission Technology Employing Optical Duobinary Coding," *IEEE J. Lightwave Technol.*, 17, 191–199, 1999.

46. K. S. Cheng and J. Conradi, "Reduction of Pulse-to-Pulse Interaction Using Alternative RZ Formats in 40 Gbit/s Systems," *IEEE Photonics Technol. Lett.*, 14, 98–100, 2002.

47. J. H. Sinsky, A. Adamiecki, C. A. Burrus, S. Chandrasekhar, J. Leuthold, and O. Wohlgemuth, "A 40-Gbit/s Integrated Balanced Optical Front End and RZ-DPSK Performance," *IEEE Photonics Technol. Lett.*, 15, 1135–1137, 2003.

48. R. A. Griffin and A. C. Carter, "Optical Differential Quadrature Phase-Shift Key (oDQPSK) for High Capacity Optical Transmission," *Proc. Optical Fiber Communication Conf.*, 2002, pp. 367–368.

49. P.- Y. Chiang, C.- C. Hsiao, and W. I. Way, "Using Precoding Techniques to Reduce the BER Penalty of an M-QAM Channel in Hybrid AM-VSB/M-QAM Subcarrier Multiplexed Lightwave Systems," *IEEE Photonics Technol. Lett.*, 10, 1177–1179, 1998.

50. L. M. Lunardi, S. Chandrasekhar, A. H. Gnauck, C. A. Burrus, and R. A. Hamm, "20-Gbit/s Monolithic p-i-n/HBT Photoreceiver Module for 1.55 μm Applications," *IEEE Photonics Technol. Lett.*, 7, 1201–1203, 1995.

51. M. Bitter, R. Bauknecht, W. Hunziker, and H. Melchior, "Monolithic InGaAs-InP p-i-n/HBT 40-Gbit/s Optical Receiver Module," *IEEE Photonics Technol. Lett.*, 12, 74–76, 2000.

52. D. Haber, M. Bitter, M. Dulk, S. Fisher, E. Gini, A. Neiger, R. Schreieck, C. Bergamaschi, and H. Jackel, " A 53 GHz Monolithically Integrated InP/InGaAs PIN/HBT Reciever OEIC with an Electrical Bandwidth of 63 GHz," *Proc. Indium Phosphide and Related Materials Conf.*, 2000, pp. 325–328.

53. K. Takahata, Y. Muramoto, H. Fukano, K. Kato, A. Kozen, O. Nakajima, and Y. Matusoka, "46.5-GHz-Bandwidth Monolithic Receiver OEIC Consisting of a Waveguide p-i-n Photodiode and a HEMT Distributed Amplifier," *IEEE Photonics Technol. Lett.*, 10, 1150–1152, 1998.

54. G. G. Mekonnen, W. Schlack, H.- G. Bach, R. Steingruber, A. Seeger, T. Engel, W. Passenberg, A. Umbach, C. Schrumm, G. Unterborsch, and S. van Waasen, "37-GHz Bandwidth InP-Based Photoreceiver OEIC Suitable for Data Rates up to 50 Gbit/s," *IEEE Photonics Technol. Lett.*, 11, 257–259, 1999.

55. Y. Zhang, C. S. Whelan, R. Leoni III, P. F. Marsh, W. E. Hoke, J. B. Hunt, C. M. Laighton, and T. E. Kazior, "40-Gbit/s OEIC on GaAs Substrate Through Metamorphic Buffer Technology," *IEEE Electron Device Lett.*, 24, 529–531, 2003.

56. P. Fay, W. Wohlmuth, A. Mahajan, C. Caneau, S. Chandrasekhar, and I. Adesida, "Low-Noise Performance of Monolithically Integrated 12-Gbit/s p-i-n/HEMT Photoreceiver for Long-Wavelength Transmission Systems," *IEEE Photonics Technol. Lett.*, 10, 713–715, 1998.

57. P. Fay, W. Wohlmuth, C. Caneau, and I. Adesida, "18.5-GHz Bandwidth Monolithic MSM/MODFET Photoreceiver for 1.55-μm Wavelength Communication Systems," *IEEE Photonics Technol. Lett.*, 8, 679–681, 1996.

58. V. Hurm, W. Benz, M. Berroth, W. Bronner, T. Fink, M. Haupt, K. Kohler, M. Ludwig, B. Raynor, and J. Rosenzweig, "10 Gbit/s Long-Wavelength Monolithic Integrated Optoelectronic Receiver Grown on GaAs," *Electronics Lett.*, 32, 391–392, 1996.

59. K. Sato, T. Hosoda, Y. Watanabe, S. Wada, Y. Iriguchi, K. Makita, A. Shono, J. Shimizu, K. Sakamoto, I. Watanabe, K. Mitamura, and M. Yamaguchi, "Record Highest Sensitivity of -28.0 dBm at 10 Gbit/s Achieved by Newly Developed Extremely-Compact Superlattice-APD Module with TIA-IC," *Proc. Optical Fiber Communication Conf.*, paper FB11-1, 2002.

60. H. Matsuda, A. Miura, Y. Okamura, H. Irie, K. Ito, T. Toyonaka, H. Takahashi, and T. Harada, "High Performance of 10-Gbit/s APD/Preamplifier Optical-Receiver Module with Compact Size," *IEEE Photonics Technol. Lett.*, 15, 278–280, 2003.

61. O. K. Tonguz and L. G. Kazovsky, "Theory of Direct-Detection Lightwave Receivers Using Optical Amplifiers," *J. Lightwave Technology*, 9, 174–181, 1991.

62. Y. K. Park and S. W. Granlund, "Optical Preamplifier Receivers: Application to Long-Haul Digital Transmission," *Optical Fiber Technology*, 1, 59–71, 1994.

63. M. Yoneyama, E. Sano, S. Yamahata, Y. Matsuoka, and M. Yaita, "17 Gbit/s pin-PD/Decision Circuit Using InP/InGaAs Double-Heterojunction Bipolar Transistors," *Electronics Lett.*, 32, 393–394, 1996.

64. M. Pfenningbauer, M. M. Strasser, M. Pauer, and P. J. Winzer, "Dependence of Optically Preamplified Receiver Sensitivity on Optical and Electrical Filter Bandwidths — Measurement and Simulation," *IEEE Photonics Technol. Lett.*, 14, 831–833, 2002.

65. B. Mason, J. M. Geary, J. M. Freund, A. Ougazzaden, C. Lentz, K. Glogovsky, G. Przybylek, L. Peticolas, R. Walters, L. Reynolds, J. Boardman, T. Kercher, M. Rader, D. Monroe, L. Ketelsen, S. Chandrasekhar, and L. L. Buhl, "40 Gbit/s Photonic Integrated Receiver with -17 dBm Sensitivity," *Proc. Optical Fiber Communication Conf.*, paper FB10-1, 2002.

7 IC Technologies for Future Lightwave Communication Systems

Taiichi Otsuji

CONTENTS

This chapter describes state-of-the-art high-speed electronic devices and IC technologies for very high-speed future lightwave communication systems. The technology of interest is for over 40-Gbit/s transmitter and receiver operations. Device technology including Si, Si-Ge, GaAs-based, and InP-based transistors as well as circuit design technology

including analog/digital/mixed-signal and optoelectronic ICs are reviewed. The speed-limiting factors are discussed to address future trends toward 100 Gbit/s and beyond.
Keywords: lightwave communication; IC; digital; analog; mixed signal; FET; BJT; HBT; HEMT

7.1 INTRODUCTION

The explosion of Internet multimedia communications has rapidly penetrated throughout the world, urgently demanding the expansion of transmission network capacity. The transmission throughput for the backbone network must be enhanced to a level of tens of terabits per second in the very near future. Figure 7.1 shows the trends in transmission throughput vs. single-channel bit rate for lightwave communication network systems at the experimental level. With the emergence of WDM (Wavelength-Division Multiplexing) and OTDM (Optical Time-Division Multiplexing) technologies, transmission throughput has, at the experimental level, exceeded 10 Tbit/s with electrical time-division multiplexing (ETDM) base rates of up to 42.7 Gbit/s [1,2]. In terms of system reliability, ease of administration, and cost, advancing the electronic integrated circuits (ICs) to achieve base rates of 40 Gbit/s and beyond is the most promising way to build practical terabit network systems.

High-speed integrated microelectronics is one of the core technologies driving the advancement of communication systems. The history of high-speed electronics, in short, is that of the progress in speed and integration of transistors [3]. The history of transistors started with the invention of field-effect transistors by J.E. Lilienfeld in 1930 and bloomed

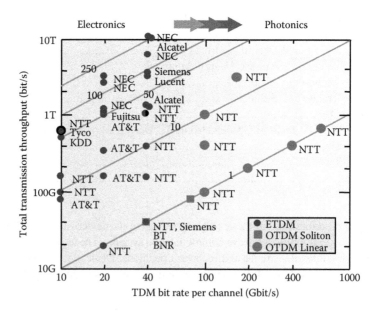

FIGURE 7.1 Transmission throughput vs. single-channel bit rate.

TABLE 7.1
State-of-the-Art Transistor Performances

Materials	Si	SiGe	GaAs			InP	
Transistor type	MOSFET	HBT	MESFET	HEMT	HBT	HEMT	HBT
Mobility (cm²/Vs)	300	500	5000	6000	5000	8500	8500
Feature size	60 nm	120 nm 20 nm	60 nm	150 nm	1.0 μm 70 nm	25 nm	800 nm 20 nm
f_T	245 GHz	350 GHz	163 GHz	100 GHz	170 GHz	562 GHz	351 GHz

with the discovery of bipolar transistors by W. Shockley, J. Bardeen, and W. Brattin in 1947. It was followed by the invention (by J. Kilby and R. Noyce) and development (by G. Moor) of ICs in 1958, MOSFETs (Metal Oxide Semiconductor FETs) by D. Kahng and M.M. Atalla in 1960, and microprocessors by Intel Corporation in 1971. In only 20 years, transistor device technology has grown into the huge industry of semiconductor-integrated electronics. So far, in accordance with Moor's scaling law (the level of integration quadruples every three years), the level of integration has been drastically elevated from IC to ULSI (ultra-large-scale integration) via LSI and VLSI. Today, complicated systems are routinely integrated on a single chip with billions of transistors.

The speed performance of transistors has been improved by more than five orders of magnitude in the last 50 years; this improvement has been based on exploration of device/process technologies including novel heterostructures, scaled semiconductor fine processes, and smart band engineering. The heterojunction bipolar transistor (HBT) was first proposed by W. Shockley in 1948, just a year after the invention of bipolar transistors. In 1957, H. Kroemer quantitatively demonstrated their superiority of speed performance [4]. T. Mimura, on the other hand, invented and developed high-electron mobility transistors (HEMTs) in 1981 [5].

Table 7.1 summarizes the performance of state-of-the-art transistors fabricated in various structures and material systems. Compound semiconductor devices appear to be the most promising for increasing the speed of electronic ICs. In particular, HBTs and HEMTs made of InP-based, GaAs-based, and SiGe-based material systems are the most attractive for breaking through the speed limit [6]. Ongoing through the device scaling, the cutoff frequencies of those transistors are now exceeding 200 GHz. At the front line of the development, 1-THz extrapolated maximum oscillating frequency has been achieved for an InP-based HBT, while over 500-GHz current-gain-cutoff frequency (f_T) has been achieved for an InP-based HEMT [7,8]. In terms of IC speed performance, 40-Gbit/s-class digital/analog ICs were extensively developed during the 1990s [9]. In late 2002, at last, K. Murata and colleagues of NTT broke through the limit of 100-Gbit/s operation by using InP-based HEMTs [10].

There have been dramatic improvements in speed not only for compound semiconductor devices but also for Si-based homo-structure transistors (MOSFETs and BJTs (bipolar junction transistors)). Recently, the record f_T of 245 GHz was achieved with a 50-nm gate-length n-channel Si MOSFET [11]. In terms of IC speed performances,

40-Gbit/s digital ICs were first developed in early 2003 by using 0.1 μm–class Si MOSFETs [12].

High-speed transistors can now offer 100-Gbit/s or 100-GHz operation for primitive digital functional ICs [10,13]. As the signal wavelength approaches the physical dimension of circuit size, however, numerous distributed parasitic effects become serious in actual IC design, which in turn critically limits the circuit speed. In such an area, digital IC design can no longer exist in the sense of "digital." Consequently, circuit advances to relax the demand on device speed are very important.

This chapter describes high-speed electronic device and IC technologies for future lightwave communication systems. The technology of interest is for greater than 40-Gbit/s transmitter and receiver operations. The topics discussed in this chapter include (1) fundamentals of lightwave communication ICs, (2) device technology, (3) fundamentals of high-speed IC design, (4) mixed-signal design approach for ultrahigh-speed ICs, and (5) future trends and technology trade-offs.

7.2 FUNDAMENTALS OF LIGHTWAVE COMMUNICATION ICS

Figure 7.2 shows basic transmitter and receiver block diagrams. The fundamental components of electronic devices are time-division multiplexer (MUX) ICs, modulator driver ICs, photodiodes, baseband amplifier ICs, decision (DEC) ICs consisting of a D-flip-flop (D-FF) for retiming and regeneration, time-division demultiplexer (DEMUX) ICs, exclusive-OR ICs for clock recovery (CR), and frequency divider (DIV) ICs for subharmonic clock generation. Simplified circuit block diagrams for MUX, DEC, CR, DEMUX, and DIV are shown in Figure 7.3.

A number of transmitters can provide high-speed lightwave communication ICs at the level of 40 Gbit/s. In particular, SiGe HBTs and InP-based HBTs/HEMTs potentially offer affordable speed performance at 40 Gbit/s. The InP-based HBTs/HEMTs also have the advantage of optoelectronic circuit integration. The fabricated device/IC performances and results for 40-Gbit/s ETDM transmission experiments are summarized in Table 7.2.

The D-FF is the critical digital circuit that limits TDM speed because it must offer clocking operation with feedback action at the highest system data rate. Figure 7.4 depicts typical high-speed D-FF circuitry. High-speed digital ICs including the D-FFs basically consist of emitter-coupled logic (ECL) for bipolar junction transistors (BJTs) or source-coupled FET logic (SCFL) for field-effect transistors (FETs). Sano analytically revealed the simplified device figure of merit f_{fom} of operating speed for those circuits [14]:

$$f_{fom} \approx 2\left[\frac{1}{f_T} + \left(\sqrt{\frac{2V_{SW}}{0.15}} + \left(0.5 + \frac{V_{SW}}{0.6} \right) \frac{f_T}{f_{max}} \right) \frac{1}{f_{max}} \right]^{-1} \quad \text{for BJTs,} \quad (7.1)$$

$$f_{fom} \approx f_T \quad \text{for FETs,} \quad (7.2)$$

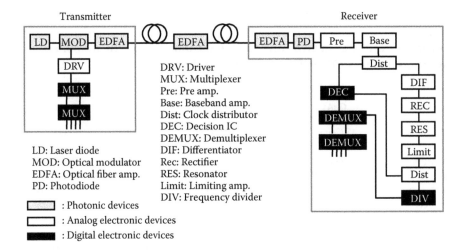

FIGURE 7.2 Basic transmitter and receiver configurations for lightwave communication systems.

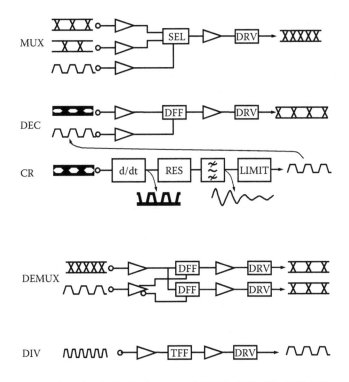

FIGURE 7.3 Simplified circuit block diagrams of MUX, DEC, CR, DEMUX, and DIV.

TABLE 7.2
Record Performance of Lightwave Communication ICs and Their Applications to Over-40-Gbit/s ETDM Transmission Experiments

| | FET/HEMT | | HBT | | |
IC	GaAs	InP	SiGe	GaAs	InP
MUX	45 Gbit/s	100 Gbit/s 70 Gbit/s[a]	50 Gbit/s	60 Gbit/[a]	50 Gbit/s
DEMUX	40 Gbit/s	50 Gbit/s[a]	60 Gbit/s	60 Gbit/s[a]	50 Gbit/s
DEC	40 Gbit/s	52 Gbit/s	40 Gbit/s	40 Gbit/s	46 Gbit/s
DIV	0–29 GHz	5–60 GHz	0–40 GHz	0–67 GHz	0–100 GHz
EX-OR		40 GHz[a]			
Modulator DRV	40 Gbit/s 6 Vpp		50 Gbit/s 4 Vpp	40 Gbit/s 2.5 Vpp	40 Gbit/s 2.3 Vpp
Baseband AMP	56 GHz	90 GHz 58 GHz[a]	55 GHz	45 GHz	85 GHz
ETDM transmission		42.7 G × 30 ch.	40 G × 273 ch.	40 G × 32 ch.	40 G × 82 ch.
exp.		3 × 125 km	1 × 117 km	3 × 100 km	3 × 100 km

[a] Results for packaged samples; others are on wafer.

where f_T, f_{max}, and V_{SW} are the current-gain cutoff frequency, the maximum oscillation frequency, and the logic swing. Figure 7.5 plots D-FF speeds vs. f_{fom}. For FETs, the maximum speed of D-FFs stays one-fourth to one-fifth of f_{fom}, while it stays one-third to one-fourth of f_{fom} for BJTs. This implies that FETs having a f_T of > 200 GHz or BJTs having balanced f_T and f_{max} of > 120 GHz are required for 40- Gbit/s D-FF operation. Advanced high-speed circuit design technology should relax the demand on transistor speed.

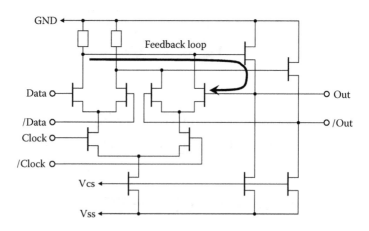

FIGURE 7.4 Basic circuit configuration of high-speed D-flip-flop.

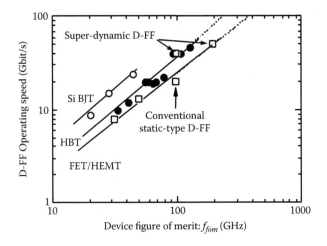

FIGURE 7.5 D-FF speed vs. device figure of merit f_{fom}.

7.3 DEVICE TECHNOLOGY

For high-speed lightwave communication ICs at the level of 40 Gbit/s, numerous transistors are available [9]. Historically, Si BJTs and MOSFETs have been part of the longest effort to improve the speed performance. However, heterostructure and/or compound semiconductor devices including Si-Ge HBTs, GaAs metal-semiconductor FETs (MESFETs), GaAs-based HBTs and HEMTs, and InP-based HBTs and HEMTs can potentially offer much higher speeds of operation than can Si BJTs and MOSFETs. Those state-of-the-art device technologies are reviewed in this section.

The speed performance of transistors is well expressed by the two figures-of-merit: the current-gain cutoff frequency (f_T) and the maximum oscillation frequency (f_{max}), which are approximately given by

$$f_T = \frac{g_m}{2\pi C_{gs(be)}},$$

(7.3)

$$f_{max} = \sqrt{\frac{f_T}{8\pi R_{g(b)} C_{gd(bc)}}},$$

(7.4)

where g_m is the transconductance, $C_{gs(be)}$ is the gate-to-source (base-to-emitter) capacitance, $C_{gd(bc)}$ is the gate-to-drain (base-to-collector) capacitance, and $R_{g(b)}$ is the gate (base) resistance. f_T is determined by the electron transit time and the CR time constant at the input node so that it may correspond to the transistor switching speed. f_{max}, on the other hand, includes an additional CR time constant due to the feedback capacitance $C_{gd(bc)}$ and the input series resistance $R_{g(b)}$, so that it may correspond well to the gain-bandwidth performance as an amplifier. Therefore, f_T/f_{max} is mainly used as a figure of merit for digital/analog ICs. Furthermore, the relation between circuit and transistor performance is slightly different between FET-based and BJT-based ICs. Since FET-based ICs can ignore the gate current, the effect of input series resistance R_g is weak. Conversely, since BJT-based ICs cannot ignore the base current, the effect of input series resistance R_b is strong. f_{max} rather than f_T, in many cases, more likely reflects the speed performance of BJT-based digital ICs.

7.3.1 Sɪ MOSFETs

Figure 7.6 depicts the cross section and the energy band diagram of Si MOSFETs. When the gate terminal is positively biased, negative charge is electrostatically induced just beneath the interface between the SiO_2 and the p-type Si, resulting in the formation of the inversion layer (the electron channel). Shortening the gate length directly reflects shortening the electron transit time, that is, the increase in the transconductance, which improves the speed performance.

The induced electronic charge is distributed in a certain thickness according to the potential gradient of the inversion layer. This causes an undesirable short-channel effect that degrades the charge controllability of the gate biasing. In order to minimize the short-channel effect, both uniformity in the electric field across the SiO_2 layer and thinner electronic charge distribution are key. Therefore, device scaling along the vertical axis (both the channel thickness and the oxide thickness) is also required to suppress the short-channel effect. Recently, a group at Toshiba developed a 60-nm gate-length n-channel MOSFET with a 1.5-nm thick ultrathin SiO_2 layer, achieving a record 245-GHz f_T [11]. As for the record IC speed performance, 40-Gbit/s MUX and DEMUX ICs have been developed by using a 120-nm gate-length Si CMOS (Complementary MOS) technology [12].

For MOSFETs with gate length of less than 100 nm, the SiO_2 layer should be as thin as a few nm; this causes a strong vertical electric field of $> 10^6$ V/cm which becomes one of the major electron-scattering factors [15]. The ultrathin oxide layer also makes the gate leakage current critical to the normal transistor operation. Those factors are the substantial limit on the speed performance of bulk MOSFETs. Recently, HfO_2, a new dielectric material having a high permittivity (6 times as high

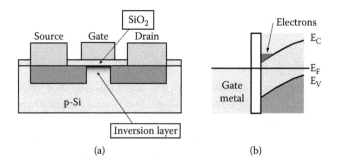

(a)   (b)

FIGURE 7.6 Si MOSFET. (a) Device cross section. (b) Energy band diagram.

as that of SiO$_2$) has been introduced to the gate insulator, which can relax the vertical electric field to about 1/6 as low as it has ever been [16]. This will assure speed improvement on device scaling down to a gate length of around 20 nm.

Silicon-on-insulator (SOI) structure and tensile-strain-induced channel structure are other typical solutions to improve MOSFET performance. The SOI structure can form a nanometer-order ultrathin channel, which can help reduce the short-channel effect. On the other hand, Takagi proposed the introduction of a SiGe base layer just under the channel [17]. Since SiGe has a lattice constant larger than that of Si, it induces tensile strain to the electrons in the channel, resulting in enhancement of the electron and hole mobility by a factor of two at most. Recently, development of sub-50-nm-class strain-induced SOI CMOS has emerged [18].

7.3.2 SiGe HBTs

Si-Ge HBTs are Si-based BJTs that replace a p-type Si base layer with a narrow-bandgap p-type Si$_{1-x}$Ge$_x$ base (see Figure 7.7). The narrow-bandgap SiGe base layer increases the discontinuity of the valence band at the emitter-base interface, which drastically reduces the injection of holes from the base to the emitter even under a highly doping condition for the base layer [19, 20]. This significantly improves the transconductance g_m as well as the parasitic base resistance R_b. As is seen in Equation 7.3 and Equation 7.4, HBTs can provide higher speed performance than that of homojunction BJTs. That is the basic concept of HBTs.

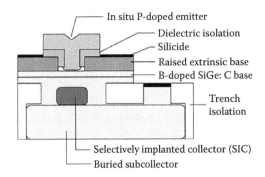

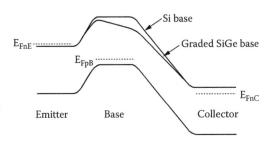

FIGURE 7.7 SiGe HBT. (a) Device cross section (after [22]). (b) Energy band diagram.

The selective epitaxial growth process, stacking the constant Ge content layer followed by the graded Ge content layer toward the emitter, can produce an extremely shallow base thickness, shortening the base transit time. Sophisticated self-aligned techniques are utilized to reduce the base resistance and collector-substrate capacitance. The record f_T of 350 GHz with an f_{max} of 107 GHz has been achieved with a 0.12×2.5-μm^2 emitter self-aligned SiGe HBT by a group at IBM [21]. Balanced high f_T and f_{max} of 207 GHz and 285 GHz have also been achieved with the same device [22]. Superior f_T and f_{max} of 122 GHz and 163 GHz have been achieved with a 0.2×1.0-μm emitter self-aligned SiGe HBT by a group at Hitachi [20]. The speed improvement of beyond 100% from Si BJTs has been obtained at the sacrifice of lowering the breakdown voltage (~2 V), which might be a serious limit for use in modulator driver ICs.

In the early stage, H.M. Rein and coworkers of Ruhr University and Siemens developed a 60-Gbit/s MUX and DEMUX ICs by using 0.3-μm emitter SiGe HBTs [23]. In the last three years, by using 0.2-μm emitter SiGe HBTs, K. Washio and coworkers of Hitachi developed a 67-GHz static 1/4 frequency divider, a 40-Gbit/s decision IC, a 40-Gbit/s 2:1 MUX IC, 1:4 DEMUX with decision IC, a 45-GHz bandwidth, and a 50-dBΩ transimpedance amplifier [20,24].

SiGe BiCMOS, the integration of high-speed SiGe HBTs and low-power CMOS, is now emerging for 40-Gbit/s-class lightwave communication LSIs. A group at IBM developed a 43-Gbit/s transceiver/receiver pair by using a SiGe BiCMOS process with an 0.18-μm emitter width [25]. The transceiver integrated a 4:1 MUX and a clock multiplexer unit while the receiver integrated a 1:4 DEMUX and a clock/data recovery. Hitachi also developed a 43-Gbit/s 16:1 MUX and 1:16 DEMUX using a similar process [26].

7.3.3   GaAs MESFETs

GaAs MESFETs are one of the most mature, mass-productive compound semiconductor transistors. By shortening the transistor feature size (the gate length) and suppressing the short-channel effects, GaAs MESFETs have been improving their speed. MESFETs in the 0.15-μm class, having both f_T and f_{max} of ~100 GHz and a transconductance (G_m) of ~500 mS/mm, were developed by NTT [27]. So far, by using 0.15-mm GaAs MESFETs, 40-Gbit/s class lightwave communication ICs have been developed. These include a 20-40-Gbit/s super-dynamic decision IC and a 44-Gbit/s 2:1 MUX IC [28].

The record f_T of 163 GHz was obtained with a 0.06-μm gate length [29]. Further shrinkage of the FETs gate length to below 0.1 μm, however, causes considerable short-channel effect. This means the digital IC speed can no longer be improved by further device scaling. Therefore, circuital advances to provide faster operation (as implemented in [27]) and/or introduction of heterostructure to make thinner electron channel beneath the gate are essential to develop $\geq$ 40-Gbit/s ICs in the real world.

7.3.4   GaAs-based and InP-based HEMTs

HEMTs are heterostructure FETs having a two-dimensional electron gas (2-DEG) channel in an intrinsic narrow-bandgap layer in between wide-bandgap carrier-supplying

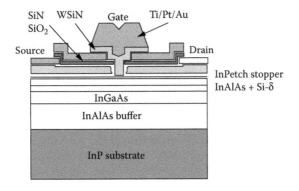

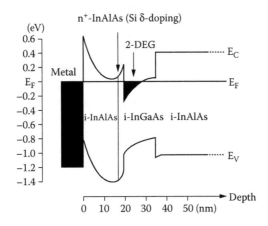

FIGURE 7.8 InP-based HEMT. (a) Device cross section. (b) Energy band diagram.

buffer layers [30]. Because of the excellent electron transport and carrier confinement property of the 2-DEG, higher-speed performance with less short- channel effect can be achieved than with standard FETs. A typical device structure and band diagram are shown in Figure 7.8.

A popular GaAs-based HEMT is the AlGaAs/InGaAs/GaAs pseudomorphic HEMT (p-HEMT). A group at Fraunhofer Institute developed a 0.15-μm gate-length p-HEMT having a f_T, f_{max}, and G_m of 100 GHz, 180 GHz, and 800 mS/mm, respectively [31]. For use in digital IC applications, 0.2-μm gate AlGaAs/GaAs/AlGaAs p-HEMTs have been utilized. They exhibit typical f_T and G_m of 60 GHz and 600 mS/mm, providing a 45-Gbit/s 2:1 multiplexer IC; 40-Gbit/s, 2.5-V_{p-p} modulator driver IC; 40-Gbit/s 1:4 demultiplexer IC; and various 20- to 30-Gbit/s class analog/digital receiver ICs [32,33]. On the other hand, Fujitsu recently developed a record 54-GHz bandwidth, 6-V_{p-p} distributed amplifier IC for a 40-Gbit/s modulator driver by using 0.15-μm gate-length InGaP/InGaAs/GaAs p-HEMTs having an f_T and G_m of 90 GHz and 525 mS/mm [34].

Over the past ten years, HEMTs consisting of the InAlAs/InGaAs modulation-doped structure on InP substrates have become not only the fastest three-terminal devices, but also uniform devices that can be applied to a wide variety of ICs for millimeter-wave and optical fiber communication systems. Compared to AlGaAs/InGaAs/GaAs p-HEMTs, InAlAs/InGaAs material systems have larger conduction-band discontinuity; this results in higher carrier confinement in the InGaAs channel (see Figure 7.8) [35]. This helps suppress the short-channel effect even for sub-100-nm gate-length devices.

Enoki and coworkers at NTT devised a highly uniform InAlAs/InGaAs/InP HEMT structure [36], which provided the opportunity to integrate complicated digital functions onto an InP-based HEMT chip. The electron channel is formed with a 15-nm thick InGaAs layer beneath a T-shaped gate with a 0.1-μm footprint. A novel InP gate-recess-etch stopper inserted into the InAlAs barrier layer dramatically improves the uniformity of transistor performance (the average threshold voltage (V_{th}); −0.65 V has a standard deviation of less than 40 mV over a 3-inch wafer), which is essential for the large-sale integration of digital logic circuits. For 0.1-μm gate HEMTs, the average transconductance (G_m), f_T, and f_{max} are 1050 mS/mm, 195 GHz, and 230 GHz, respectively. Progress on shortening the gate length has recently been made by groups at NTT and Fujitsu [8, 37, 38]. The record f_T of 562 GHz has been realized with a 25-nm-gate length by Yamashita and coworkers at Fujitsu [8].

By using 0.1-μm gate HEMTs, NTT developed 40- to 100-Gbit/s class transmitter and receiver ICs of various types, which included the record 100-Gbit/s 2:1 MUX IC, a 10-dB, 90-GHz baseband amplifier IC, a 50-Gbit/s super-dynamic decision IC, a 50-Gbit/s 1:2 DEMUX IC, and a 43-Gbit/s clock and data recovery IC [10,39–42]. Recently, Fujitsu also developed a 100-Gbit/s 2:1 multiplexer IC and a 43-Gbit/s 4:1 multiplexer IC containing a 52-Gbit/s static D-flip-flop by using 0.13-μm gate HEMTs [43,44].

7.3.5 GaAs-based and InP-based HBTs

Figure 7.9 shows a typical device cross section and energy band diagram for InP-based HBTs. The concept of introducing the heterostructure is basically the same as that for SiGe HBTs. Compared to the SiGe HBTs, in which a narrower bandgap base is introduced, GaAs-based and InP-based HBTs basically introduce a wider bandgap emitter.

AlGaAs/GaAs material systems are widely accepted for GaAs-based HBTs and were the first to enable 40-Gbit/s D-flip-flop IC operations [45]. Over the past 15 years, the AlGaAs/GaAs HBTs have improved their speed performance by incorporating smart device structures such as a novel quasi-ballistic transport mechanism [46]. T. Ishibashi and coworkers at NTT developed ballistic collection transistors (BCT's) having a f_T and f_{max} of 170 GHz and 190 GHz with 2-μm emitter size at a collector current density (J_c) of 50 kA/cm² [47]. However, due to serious reliability issues under high J_c, the high-speed performance has not been reflected during practical IC productions. Runge and coworkers at Rockwell developed 20- to 40-Gbit/s class analog/digital ICs, including a 40-Gbit/s 4:1 MUX IC and a 30-Gbit/s 1:4 DEMUX IC, by using standard 0.13-mm emitter AlGaAs/GaAs HBTs [48].

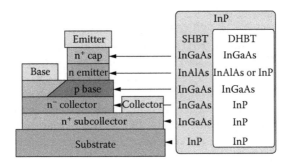

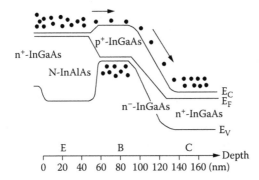

FIGURE 7.9 InP-based HBT. (a) Device cross section. (b) Energy band diagram.

AlGaAs/InGaAs HBTs with a p+ regrown GaAs extrinsic base layer are the other type of GaAs-based HBTs. NEC developed a 40-Gbit/s receiver chip set by using 1.6-μm emitter HBTs having an f_T and f_{max} of 110 and 270 GHz, respectively [49].

InGaP/GaAs HBTs have attracted much attention due to their superior speed and reliability. The large valence-band discontinuity at the InGaP-emitter/GaAs-base interface improves the carrier injection efficiency compared with the AlGaAs/GaAs interfaces [50]. Significant improvements in long-term reliability have been reported by replacing the AlGaAs emitter material with InGaP, which assured an MTTF of >10^5 hours under relatively high J_c of 50 kA/cm^2 [51,52]. Researchers at Hitachi developed a 39.5-GHz static frequency divider IC by using 0.5-μm emitter InGaP/GaAs HBTs having an f_T and f_{max} of 156 and 255 GHz, respectively [53]. A group at Yokogawa developed 50-Gbit/s modulator driver and 2:1 multiplexer ICs based on InGaP/InGaAs HBTs having a f_T and f_{max} of around 170 GHz [54].

InP-based HBTs can potentially offer the greatest improvement in speed over currently available transistors, with an excellent f_{max} of 1000 GHz and f_T of 350 GHz; these values are more than three times as high and twice as high, respectively, as those for GaAs devices [7,55]. In the early stage, the group at NTT improved the speeds of InP-based HBTs by incorporating the self-aligned

InP/InGaAs composite collector structure with a hexagonal emitter geometry, providing an f_T and f_{max} of 228 GHz and 227 GHz, respectively [56]. By using more reliable carbon-doped HBTs having relatively low f_T and f_{max} of 116 and 169 GHz, the NTT group has developed various high-speed digital ICs including a 40-Gbit/s decision IC and a 1:4 demultiplexer IC [57–59]. Recently, Ida and coworkers at NTT achieved the record f_T of 351 GHz and an f_{max} of 288 GHz with a self-aligned DHBTs having a 20-nm thick ultra-thin graded InGaAs base and a 0.8-μm-wide step-graded InGaAs/InGaAsP/InP emitter [55]. A balanced high f_T and f_{max} of 329 GHz has also been achieved by a similar DHBT having a 30-nm-thick base layer [55].

By using the novel substrate transfer process that allows very narrow collector stripes, Rodwell and coworkers at UCSB raised the f_{max} value to 1.08 THz [7]. The transferred-substrate HBTs were used to realize various analog/digital ICs over 40-Gbit/s, including an 85-GHz distributed amplifier IC and a 75-GHz static frequency divider IC [60,61]. Key technology for those HBTs is the device structure that can effectively reduce the parasitic base resistance and the base-collector capacitance, which directly reflects on improvement of the f_{max}.

The production-level AlInAs/InGaAs single-HBTs have also gained speed performance at the level of 130–250-GHz f_T and f_{max} [62]. They were used by HRL to produce the record 72.8-GHz 1/2 static frequency divider IC, and by Bell Laboratories and HRL to produce a 50-Gbit/s 4:1 MUX IC [63,64]. Very recently, HRL succeeded in the development of a 100-GHz 1/2 static frequency divider IC by using their upgraded DHBT technology [13]. Researchers at OPTO+ developed a 40-Gbit/s 2:1 MUX IC and master-slave D-flip-flop IC using InGaAs/InGaAsP/InP double HBTs having f_T and f_{max} of up to 160 and 220 GHz, respectively [65,66]. A group at UCSB and IQE has recently yielded the fastest production-level DHBTs exhibiting f_T and f_{max} of 282 GHz and $\geq$ 450 GHz [67].

7.3.6 Optoelectronic ICs

Photoreceivers that monolithically integrate a pin-photodiode and a transimpedance amplifier are the fundamental optoelectronic ICs (OEICs) for lightwave communication ICs. So far, > 40-GHz bandwidth photoreceiver OEICs have been developed by several research groups using InP/InGaAs pin-PD-DHBTs, GaAs pin-PD-PHEMTs, InP/InGaAs pin-PD-HBTs, and InAlAs/InGaAs/InP waveguide- pin-PD HEMTs [68–71]. One of the big advantages of those OEICs is the excellent interconnection-dependent parasitic-free characteristic that becomes very difficult to achieve in the high-speed region of 40 Gbit/s and beyond.

Recently, a new type of wide-band, high-output-saturation characteristic InP/InGaAs photodiode, called a uni-traveling-carrier photodiode (UTC-PD), was realized by Ishibashi and coworkers at NTT [72]. Its extremely high output saturation voltage can eliminate transimpedance and/or baseband amplifiers and can directly drive decision ICs under a 50-Ω load impedance condition. Such an optical receiver was monolithically integrated using a UTC-PD and a HEMT super-dynamic decision IC, and error-free 40-Gbit/s return-to-zero-mode receiver operation was successfully demonstrated [73].

7.4 FUNDAMENTALS OF HIGH-SPEED DIGITAL IC DESIGN [92]

A typical flowchart for high-speed digital IC design is shown in Figure 7.10. It consists of model parameters extraction for active/passive circuit elements, circuit simulation, layout design, modeling parasitic elements, and feedback to the circuit design. For digital IC design, a HSPICE™ simulator is frequently used. The accuracy of the simulation relies predominantly upon the accuracy of the circuit model parameters.

The large signal equivalent circuit for a transistor includes more than 30 DC/AC parameters. The HSPICE circuit simulator and the ADS™ microwave design simulator are utilized for parameter extraction. All of them are satisfactorily extracted according to the following well-considered but complicated procedure:

1. Optimizing the bias-independent DC parameters
2. Optimizing the large-signal DC parameters (the gate diode, the drain current as functions of biases)
3. Optimizing the small-signal AC parameters at each bias point
4. Optimizing the large-signal AC parameters (bias-dependence of junction capacitances)

In steps 1, 3, and 4, the parameters are optimized by fitting the simulated S parameters of the equivalent circuit to the measured ones.

The most important passive elements to be considered are interconnection lines and inductances. The simple way for those expressions is to obtain their frequency responses as sets of S parameters. However, they do not work in HSPICE, so they are modeled as L-C-R networks by fitting their frequency responses to those obtained from the electromagnetic simulation for modeled structures. All the critical paths should be modeled.

According to the procedure mentioned above, a super-dynamic decision IC was designed and fabricated at NTT [74,75]. Typical measured and simulated waveforms for PN-7 stage pseudorandom data input are shown in Figure 7.11. At 40 Gbit/s in

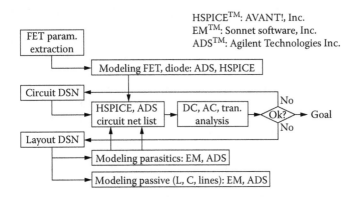

FIGURE 7.10 Design flowchart for high-speed digital ICs. (After [92].)

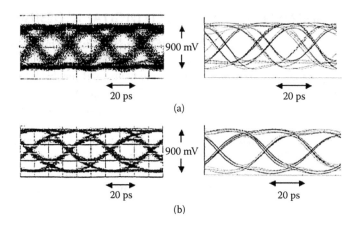

FIGURE 7.11 Output waveforms of a decision IC for PN-7 pseudorandom data pattern input. Left: measured on wafer. Right: simulated. (a) At 40 Gbit/s. (b) At 36 Gbit/s. (After [92].)

(a), the eye opening is severely degraded due to inter-bit distortion. When the data rate goes down to 36 Gbit/s in (b), the jitter is sufficiently suppressed. The simulated results perfectly express the reality.

There are important issues for modeling the HBTs. For accurate modeling, we must consider the following physical phenomena. Standard BJT models in HSPICE™ have some limitations: (i) high injection effect, which includes a complicated mixture of the Kirk effect, emitter-crowding effect, and thermal effect; (ii) the velocity overshoot, significantly contributing to the slow collector transit time; and (iii) thinner collector structure, which causes the bias dependence of the base-collector capacitance C_{bc} to be nonuniform. Designers, therefore, should take the base-collector bias (V_{bc}) dependence of C_{bc} into account. Large variation in C_{bc} at the low reverse-biased V_{bc} region would cause dispersive switching, resulting in severe timing jitter.

7.5   MIXED-SIGNAL ISSUES IN HIGH-SPEED IC DESIGN [92]

Figure 7.12 shows the IC design criteria in terms of wavelength vs. circuit size. As data rates increase to 40 Gbit/s and beyond, the wavelength of the signal approaches the physical size of IC chips, which gives rise to substantial difficulties in designing digital ICs, such as (i) numerous parasitics, (ii) cavity resonance, (iii) bandwidth limitation due to lumped-circuit treatment, and (iv) ultra-broadband operation from near-DC to the maximum bit rate. In such an area, digital IC design can no longer exist in the sense of "digital" but in an "analog" or "mixed signal" fashion. Key technologies in high-speed IC design are (i) optimizing the process design, (ii) minimizing the parasitic capacitance by optimizing operating bias conditions and/or by introducing distributed treatment, and (iii) minimizing the effective logic swing to accelerate transitions by introducing dynamic operation. Issue (i) is a process design issue, while (ii) and (iii) are the circuit design issues.

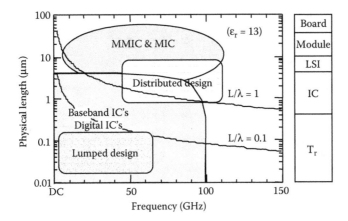

FIGURE 7.12 IC design criterion in terms of wavelength vs. circuit size.

7.5.1 DISTRIBUTED DESIGN APPROACH

Very high-speed digital ICs are generally based on SCFL or ECL series-gated circuitry. The input node is the high impedance (in the order of tens of kΩ) gate/base, while the output is the low impedance ($1/(g_m + g_{ds})$, where g_m is the transconductance and g_{ds} is the drain conductance) source/emitter follower. When the line length approaches one tenth of the signal wavelength (280 μm for ε_r of 7.0 at 40 GHz), the mismatch between the line impedance (Z_L) and the output impedance causes severe multiple reflections.

As shown in Figure 7.13, single- or double-railed strip lines (SLs) are generally used for the digital IC layout. However, Z_L of the SL is relatively high (200 Ω). The double-railed SLs help reduce the Z_L, but still larger than the SCFL output impedance, and largely varies with surrounding environments because of the lack of steady ground. Therefore, severe distortion occurs and degrades speed performance.

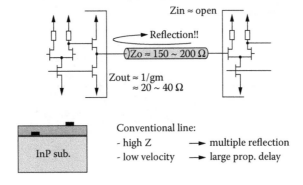

FIGURE 7.13 Circuit diagram and cross-sectional structure for conventional interconnection with striplines. (After [92].)

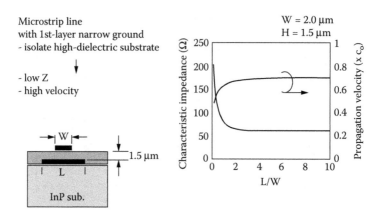

FIGURE 7.14 Impedance-controlled microstrip lines: cross-sectional structure and electrical property. (After [92].)

A second-layer microstrip line (MSL) with a first-layer ground, shown in Figure 7.14, is best suited for high-speed, low-impedance signal transmission on ICs. The Z_L of the MSL can be lowered to match the SCFL output impedance (20–40 Ω). Its structure, isolated from the substrate, makes the effective dielectric constant very low (2.6), and increases the propagation velocity by 50%. This makes for better matching to the output impedances of the SCFLs. A good example of this is an 80-Gbit/s multiplexer IC [76] as shown in Figure 7.15. A high-speed, imped-ance-matched transmission design was introduced for intercell connection of critical paths of 400-μm long. This drastically reduced the waveform distortion due to multiple reflections, resulting in a speed improvement of 30% over the conventional lumped circuit design.

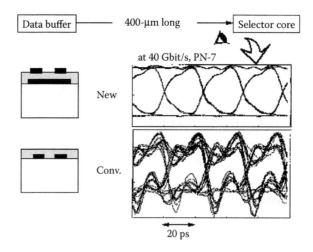

FIGURE 7.15 Simulated 40-Gbit/s eye diagrams at the selector core input. Upper: MSL interconnection. Lower: conventional SL interconnection. (After [76].)

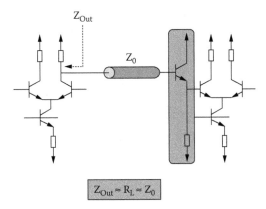

FIGURE 7.16 Another option of impedance-matched interconnection for ECL circuits. (After [92].)

In case of ECLs, the output impedance is extremely low ($< 10\ \Omega$), so that one may choose another option where the intercell connection is made in between the collector node of the emitter-coupled differential pair and the emitter-follower input. Z_L, thus, becomes practically high so as to match the load impedance Z_{OUT} (see Figure 7.16).

7.5.2 Dynamic Response Design Approach

This approach is unique for digital IC designs based on large-signal, time-domain response. A typical example is the super-dynamic D-FF (see Figure 7.17) [74,75]. The circuit features (i) a smaller latching current than the reading current, and (ii) a source-coupled negative feedback pair, which drastically reduces the effective logic swing while keeping a fast signal transition slew rate. The circuit can operate about twice as fast as the conventional master-slave D-FF's can. Up to 50-Gbit/s DEC operation has been achieved using 0.1-μm InP-based HEMTs [40]. The basic idea of dynamical logic swing shrinkage to accelerate the switching speed was initially implemented on a D-FF by Murata et al. of NTT, and Ishii et al. of NTT [77,78]. So far, various types of D-FF circuits have been proposed using a similar design [24,79,80].

Key ideas in designing DEMUXs are (i) wider clock phase margin and (ii) higher input sensitivity, which is due to the performance of the first-stage D-FF and to the clock-skew distribution. All the design techniques mentioned above reflect these key ideas.

7.5.3 Technology Trade-offs

In this subsection, the technology trade-offs among the above mentioned transistor types (HBTs vs. HEMTs) are discussed from the viewpoint of high-speed, low-power, digital circuit integration. Sokolich at HRL presented the power efficiency of the flip-flop speed for various process technologies [81], which is shown in Figure 7.18 along with additional results for NTT's InP-based HEMT ICs. As shown

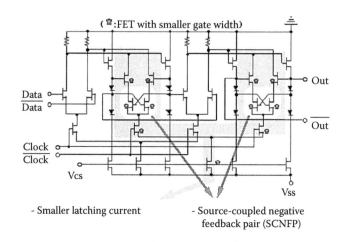

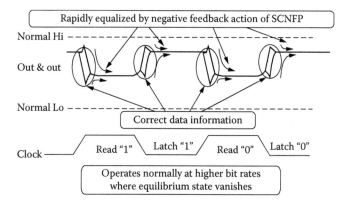

FIGURE 7.17 Super-dynamic D-FF. (a) Circuit diagram. (b) Principle of high-speed operation. (After [74].)

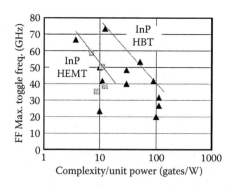

FIGURE 7.18 Power efficiency of the flip-flop speed for various transistors: for HEMTs, for HBTs. (After [81].)

in the figure, HBTs can enable operation at a power almost one order of magnitude lower than HEMTs.

According to Sano's analytical expressions for the device figure of merit f_{fom} (see Equation 7.1 and Equation 7.2), it is noted that the ECLs can accelerate the switching speed by reducing the logic swing, but the SCFLs cannot. This advantage for ECLs is mainly due to the rapid increase in f_T (or G_m) with increase in the base bias voltage, resulting in low-power, high-speed operation. As a result, in terms of the speed vs. logic swing, HBTs can run faster with lower swing down to 300 or 250 mV, but it is very hard to do this with HEMTs. As mentioned in Section 7.4, however, designers should take the V_{bc} dependence of the collector-base capacitance C_{bc} into account when shrinking the logic swing.

In terms of large-scale integration, HBTs are limited by thermal density, while HEMTs are limited by actual physical density. Thus, HBT ICs must be carefully integrated. In terms of clear eye opening, HBTs need extra care for the complexity of carrier motions. HEMTs have the significant advantage of easy design due to their simple unipolar actions. One straightforward consequence may suggest that HEMT ICs are better suited for transmitter IC design, while HBTs are better suited for receiver IC design.

7.6 FUTURE TRENDS AND SUBJECTS

7.6.1 Device Scaling

"How much faster can transistors and ICs operate?" For scaled Si-Ge HBTs, superior speed performance of 350-GHz f_T has already been realized with an ultra-thin ($\approx$ 10 nm) base layer [21]. As mentioned in Section 7.4.1, device scaling sacrifices the breakdown property. Figure 7.19 plots the breakdown voltages vs. current-gain cutoff frequencies f_T for various transistors. Compared to the SiGe HBTs, III-V compound devices, in particular for InP-based DHBTs, have superior breakdown property. This weak breakdown property becomes a critical limiting factor for speed improvement.

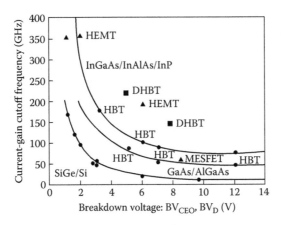

FIGURE 7.19 Transistor cutoff frequency vs. breakdown voltage for various transistors. (After [92].)

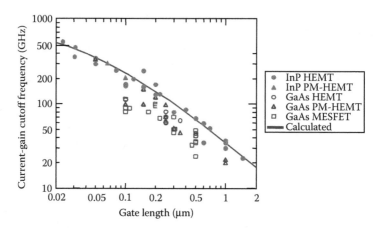

FIGURE 7.20 Transistor current-gain cutoff frequency vs. gate length for FETs and HEMTs. (add data to [83].)

Another problem for SiGe HBTs is the parasitic effects due to the Si-substrate conductivity [82], which becomes much more serious than the effects for III-V devices with semi-insulating substrates. Therefore, reduction of the substrate effects becomes a critical issue. The transferred substrate or SOI (silicon on insulator) structure will be a solution to this problem [82].

As to the transferred substrate InP-based HBTs, the f_{max} values can be improved directly, but the structure does not affect the base transit time and the base-emitter capacitance. This means the f_T values cannot be improved. As mentioned in Section 7.2, in order to efficiently improve the IC speed, an increase in f_T value is also necessary.

Figure 7.20 plots the trend in f_T vs. gate length for various FETs. In the shorter gate-length region of < 0.1 μm, the f_T curve tends to be saturated. Enoki et al. predicted that the speed performance limit of scaled InP-based HEMTs is around 700 GHz due to tunneling leakage current and parasitic capacitance [83]. Figure 7.21 plots the contour map of the SCFL inverter gate delay time (t_{pd}) on a $f_T - g_m$ plane. The vertical axis corresponds to the device scaling along the vertical axis (the barrier layer thickness) while the horizontal axis corresponds to the lateral scaling (the gate length). Thinning the barrier is limited by the punch-through phenomenon due to quantum-mechanical tunneling, while shortening the gate length makes the unscaled gate-fringing capacitance significant. Figure 7.21 suggests that the speed limit exists around 2 ps of t_{pd} when conventional HEMT structures with InP-based material systems are assumed. This implies that the speed limit of D-FF operation stays around 100 Gbit/s and that of MUX operation stays around 200 Gbit/s. Further speed improvement will require breakthroughs such as new material systems and/or new device structures having better electron transport properties.

Matloubian et al. of HRL have proposed an antimonide-based wider bandgap and larger conduction-band discontinuity material systems like InGaAs/AlGaPSb lattice-matched to an InP substrate, which may lead to terahertz operation for HEMTs with a high breakdown voltage [84]. Bolognesi et al. of Simon Fraser University

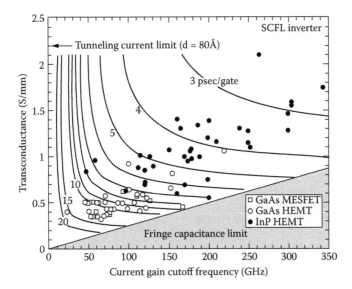

FIGURE 7.21 Correlation of FET-based inverter logic gate delay time with transistor performance. (After [83].)

introduced a GaAsSb material for the base layer in place of InAlAs for InP-based DHBTs, exhibiting over 200-GHz f_T and f_{max} with a 1-μm emitter width [85]. For both HEMTs and HBTs, larger conduction or valence band discontinuity can produce higher carrier confinement, resulting in higher-speed operation for scaled devices. For several years, antimonide-based compound semiconductors (ABCS) have been extensively investigated as the key material systems of the next generation of higher-speed transistors [86].

7.6.2 NEW ASPECTS OF OPTOELECTRONIC IC DESIGN APPROACHES

The above discussions lead to limiting the cutoff frequencies of InP-based HBTs and HEMTs at around terahertz. The trends also show that scaled InP-based HBTs or HEMTs can potentially offer 100-Gbit/s static D-FF operation and 200-Gbit/s MUX operation. Is it feasible? In such an ultrahigh-speed region, however, numerous parasitic effects such as interconnection delay critically limit the circuit speed. In addition, conventional electrical I/O interfaces do not work anymore because allowable interconnection lengths for propagating 160-Gbit/s data via metallic coaxial cables would be as short as 10 cm at best.

One direction is minimizing parasitics and power consumption to make sense of lumped circuit designs. Optoelectronic integration with highly functional devices will be one solution. A resonant tunneling diode (RTD) is a highly functional quantum effect device that exhibits a negative differential resistance yielding nonlinear, bistable operation. Maezawa devised monostable-bistable transition logic elements (MOBILEs) consisting of a serial connection of two RTDs, one for a load

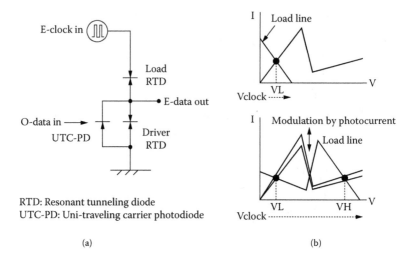

(a)   (b)

FIGURE 7.22 Optoelectronic D-FF consisting of a MOBILE (two resonant tunneling diodes) and a uni-traveling-carrier photodiode. (a) Circuit diagram. (b) Operation principle. Upper: when the clock voltage is low. Lower: when the clock voltage is high.

and the other for a driver (see Figure 7.22) [87]. The MOBILE can perform data latching operation equivalent to the D-FF function. The logic state can be controlled by modulating the drive current of the driver RTD. Sano introduced an optoelectronic scheme for modulating the driver, which is the monolithic integration of only one UTC-PD and a MOBILE on an InP substrate (see Figure 7.22). This OEIC has demonstrated the direct demultiplexing of optical 80-Gbit/s data to electrical 40-Gbit/s data and decision operation for 80-Gbit/s RZ data [88,89]. An optoelectronic injection-locked oscillator integrating one UTC-PD and one RTD successfully extracted a subharmonic 11-GHz electrical clock from an optical 44-Gbit/s data input [90].

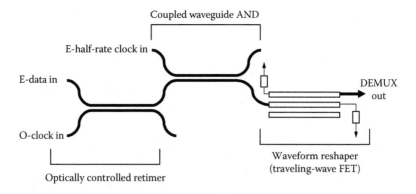

FIGURE 7.23 Ultrafast optoelectronic traveling-wave demultiplexer. (After [93].)

As an alternative approach, traveling-wave or distributed circuit designs should also be considered. One unique concept transforms a sequential (feedback) FF circuit into an equivalent feed-forward circuit by making full use of distributed design topology and the management of the propagation modes of electromagnetic waves [91]. Narahara et al. of NTT proposed a unique ultrafast optoelectronic traveling-wave demultiplexer [93], as shown in Figure 7.23. The first stage bunches the electrical data output timing to the optical clock timing by means of permittivity modulation due to photoexcited carriers. The second stage consists of a coupled waveguide that performs the demultiplexing operation with respect to the electrical half-rate clock. The final stage is the traveling-wave FET which reshapes the demultiplexed data signals. All the functions are performed in a feed-forward fashion. The speed limit of this type of the circuit has been estimated to far exceed 100 Gbit/s.

7.7 CONCLUSIONS

In this chapter, recent advances in high-speed electronic device and IC technologies for very high-speed lightwave communication systems were reviewed. Field tests of 40-Gbit/s ETDM have been performed in Europe, the United States, and Japan using SiGe HBTs, InP HBTs, and InP HEMTs. Si-Ge HBTs are attractive for mass-productive, low-power, highly-integrated applications at around 40 Gbit/s. Scaled InP-based HBTs and HEMTs can potentially offer the highest speed performance of all the transistors so that 100-Gbit/s logic ICs will be realized in the very near future. New material systems including antimonide-based compound semiconductors would enable further extension of the transistor speed limit. There exist numerous parasitics both inside and outside the chips, which critically limits the IC speed. Smart optoelectronic integration with highly functional devices will be a possible solution for an over-100-Gbit/s world.

ACKNOWLEDGMENTS

The author's contributions to the development of IC fabrications were made when he was at NTT Laboratories. The author would like to acknowledge all the researchers who were involved in this work for their extensive contributions and valuable discussion.

REFERENCES

1. K. Fukuchi, K. Kasamatsu, M. Morie, R. Ohhira, T. Ito, K. Sekiya, D. Ogawahara, and T. Ono, "10.92-Tb/s (273 x 40-Gb/s) triple-band/ultra-dense WDM optical-repeated transmission experiment," *Tech. Dig. Opt. Fiber Comm. Conf.*, PD24–1–4, 2001.
2. S. Bigo, Y. Fregnac, G. Chorlet, S. Borne, P. Tran, C. Simonneou, D. Bayart, A. Joudan, J.-P. Hamaide, W. Idler, R. Dischler, G. Veith, H. Gross, and W. Poehlmannk, "10.2 Tbit/s (256 × 42.7 Gbit/s PDM/WDM) transmission over 100 km TelaLight fiber with 1.28 bit/s/Hz spectral efficiency," *Tech. Dig. Opt. Fiber Comm. Conf.*, PD25–1–4, 2001.

3. R.M. Warner, "Microelectronics: its unusual origin and personality," *IEEE Trans. Electron. Devices*, 48, 2457–2467, 2001.

4. H. Kroemer, "Quasi-Electric and Quasi-Magnetic Fields in Non-Uniform Semiconductors," *RCA Rev.*, 18, 332–342, 1957.

5. T. Mimura, S. Hiyamizu, T. Fujii, and K. Nanbu, "A new field-effect transistor with selectively doped GaAs–n-Al$_x$Ga$_{1-x}$As heterojunctions," *Jpn. J. Appl. Phys.*, 19, L225–L227, 1980.

6. M. Rodwell, Ed., *Selected Topics in Electronics and Systems* Vol. 21, High Speed Integrated Circuit Technology: Towards 100 GHz Logic, World Scientific Publishing, Singapore, 2001.

7. Q. Lee, S.C. Martin, D. Mensa, R.P. Smith, J. Guthrie, S. Jaganathan, Y. Betser, T. Mathew, S. Krishnan, L. Samoska, and M.J.W. Rodwell, "Submicron transferred-substrate heterojunction bipolar transistors with greater than 1 THz f$_{max}$," postdeadline paper, IEEE Dev. Res. Conf., 1999.

8. Y. Yamashita, A. Endoh, K. Shinohara, K. Hikosaka, T. Matsui, S. Hiyamizu, and T. Mimura, "Pseudomorphic In$_{0.52}$Al$_{0.48}$As/In$_{0.7}$Ga$_{0.3}$As HEMTs with an ultrahigh f$_T$ of 562 GHz," *IEEE Electron Devices Lett.*, 23, 573–575, 2002.

9. Keh-Chung Wang, Ed., *Selected Topics in Electronics and Systems*, Vol. 13, High-Speed Circuits for Lightwave Communications, World Scientific Publishing, Singapore, 1999.

10. K. Murata, K. Sano, H. Kitabayashi, S. Sugitani, H. Sugahara, and T. Enoki, "100-Gbit/s logic IC using 0.1-μm-gate-length InAlAs/InGaAs/InP HEMTs," *Dig. IEEE Int. Electron Device Meet.*, 937–939, 2002.

11. H.S. Momose, E. Morifuji, T. Yoshitomi, T. Ohguro, M. Saito, and Hiroshi Iwai, "Cutoff Frequency and Propagation Delay Time of 1.5-nm Gate Oxide CMOS," *IEEE Trans. Electron Devices*, 48, 1165–1174, 2001.

12. D. Kehrer, H-D. Wohlmuth, H. Knapp, M. Wurzer, A. Scholtz, "40Gb/s 2:1 Multiplexer and 1:2 Demultiplexer in 120nm CMOS," *Dig. Tech. Papers IEEE Int. Solid State Circuits Conf.*, paper no. 19.6, 2003.

13. M. Mokhtari, C. Fields, and R.D. Rajavel, "100 + GHz static divide-by-2 circuit in InP-DHBT technology," *Tech. Dig. IEEE GaAs IC Symp.*, 291–293, 2002.

14. E. Sano, Y. Matsuoka, and T. Ishibashi, "Device figure-of-merit for high-speed digital ICs and baseband amplifiers," *IEICE Trans. Electron.*, E78-C, 1182–1188, 1995.

15. S. Villa, A.L. Lacaita, L.M. Perron, and R. Bez, "A Physically-Based Model of the Effective Mobility in Heavily-Doped n-MOSFETs," *IEEE Trans. Electron Dev.*, 45, 110–115, 1998.

16. B. Tavel, X. Garros, T. Skotnicki, F. Martin, C. Leroux, D. Bensahel, M.N. Semeria, Y. Morand, J.F. Damlencourt, S. Descombes, F. Leverd, Y. Le-Friec, P. Leduc, M. Rivoire, S. Jullian, and R. Pante, "High performance 40 nm nMOSFETs with HfO$_2$ gate dielectric and polysilicon damascene gate," *Dig. IEEE Int. Electron Dev. Meet.*, 429–432, 2002.

17. S. Takgai, J.L. Hoyt, J.J. Welser, and J.F. Gibbons, "Comparative study of phonon-limited mobility of two-dimensional electrons in strained and unstrained Si metal–oxide–semiconductor field-effect transistors," *J. Appl. Phys.*, 80, 1567–1577 (1996).

18. T. Mizuno, N. Sugiyama, T. Tezuka, T. Numata, T. Maeda, and S. Takagi, "Design for scaled thin film strained-SOI CMOS devices with higher carrier mobility," *Dig. IEEE Int. Electron Device Meet.*, 31–34, 2002.

19. D.L. Harame, D.C. Ahlgren, D.D. Coolbaugh, J.S. Dunn, G.G. Freeman, J.D. Gillis, R.A. Groves, G.N. Hendersen, R.A. Johnson, A.J. Joseph, S. Subbanna, A.M. Victor, K.M. Watson, C.S. Webster, and P.J. Zampardi, "Current status and future trends of SiGe BiCMOS technology," *IEEE Trans. Electron Devices*, 48, 2575–2594, 2001.

20. K. Washio, "Self-aligned Si BJT/SiGe HBT technology and its application to high-speed circuits," *Int. J. High Speed Electron. Sys.* 11, 77–114, 2001.

21. J.-S. Rieh, B. Jagannathan, H. Chen, K.T. Schonenberg, D. Angell, A. Chinthakindi, J. Florkey, F. Golan, D. Greenberg, S.-J. Jeng, M. Khater, F. Pagette, C. Schnabel, P. Smith, A. Stricker, K. Vaed, R. Volant, D. Ahlgren, and G. Freeman, "SiGe HBTs with cut-off frequency of 350 GHz," *Dig. IEEE Int. Electron Device Meet.*, 771–774, 2002.

22. B. Jagannathan, M. Khater, F. Pagette, J.-S. Rieh, D. Angell, H. Chen, J. Florkey, F. Golan, D.R. Greenberg, R. Groves, S.J. Jeng, J. Johnson, E. Mengistu, K.T. Schonenberg, C.M. Schnabel, P. Smith, A. Stricker, D. Ahlgren, G. Freeman, K. Stein, and S. Subbanna, "Self-aligned SiGe NPN transistors with 285 GHz f_{max} and 207 GHz f_T in a manufacturable technology," *IEEE Electron Device Lett.*, 23, 258–260, 2002.

23. H.-M. Rein, "Si and SiGe bipolar ICs for 10 to 40 Gb/s optical-fiber TDM links," *Int. J. High Speed Electron. Sys.*, 9, 347–383, 1998.

24. K. Washio, R. Hayami, E. Ohue, K. Oda, M. Tanabe, H. Shimamoto, and M. Kondo, "67-GHz static frequency divider using 0.2-µm self-aligned SiGe HBT," *IEEE Trans. Microwave Theory Tech.*, 49, 3–8, 2001.

25. M. Meghelli, A. Rylyakov, S. Zier, and M. Sorna, "A 0.18 µm SiGe BiCMOS Receiver and Transmitter Chipset for SONET OC-768 Transmission Systems," *Dig. Tech. Pap. IEEE Int. Solid State Circuits Conf.*, paper no. 13.1, 2003.

26. Koyama, T. Harada, H. Yamashita, R. Takeyari, N. Shiramizu, K. Ishikawa, M. Ito, S. Suzuki, T. Yamashita, S. Yabuki, H. Ando, T. Aida, K. Watanabe, K. Ohhata, S. Takeuchi, H. Chiba, A. Ito, H. Yoshida, A. Kubota, T. Takahishi, and H. Nii, "43Gb/s Full-Rate-Clock 16:1 Multiplexer and 1:16 Demultiplexer with SFI-5 Interface in SiGe BiCMOS Technology," *Dig. Tech. Pap. IEEE Int. Solid State Circuits Conf.*, paper no. 13.2, 2003.

27. K. Nishimura, K. Onodera, S. Aoyama, M. Tokumitsu, and K. Yamasaki, "High performance 0.1 µm self-aligned-gate GaAs MESFET technology," *Proc. European Solid-State Device Res. Conf. (ESSDERC)'96*, 865–868, 1996.

28. T. Otsuji, K. Murata, K. Narahara, K. Sano, and E. Sano, "20-40-Gbit/s-class GaAs MESFET digital ICs for future optical fiber communications systems," *Int. J. High Speed Electron. Sys.*, 9, 399–435, 1998.

29. M. Tokumitsu, M. Hirano, T. Otsuji, S. Yamaguchi, and K. Yamasaki, "A 0.1-µm self-aligned-gate GaAs MESFET with multilayer interconnection structure for ultra-high-speed ICs," *Tech. Dig. Int. Electron Device Meeting (IEDM)*, 211–214, 1996.

30. T. Mimura, S. Hiyamizu, T. Fujii, and K. Nanbu, "A new field-effect transistor with selectively doped GaAs–n-$Al_xGa_{1-x}As$ heterojunctions," *Jpn. J. Appl. Phys.*, 19, L225–L227, 1980.

31. M. Schlechtweg, W.H. Haydl, A. Bangert, J. Braustein, P.J. Tasker, L. Verweyer, H. Massler, W. Bronner, A. Hulsmann, and K. Kohler, "Coplanar millimeter-wave ICs for W-band applications using 0.15 µm pseudomorphic MODFETs," *IEEE J. Solid State Circuits*, 31, 1426–1434, 1996.

32. Z. Lao, M. Lang, V. Hurm, Z. Wang, A. Thiede, M. Schlechtweg, W. Bronner, G. Kaufel, K. Kohler, A. Hulsmann, B. Raynor, and T. Jakobus, "20-40-Gbit/s GaAs-HEMT chip set for optical data receiver," *Int. J. High Speed Electron. Sys.*, 9, 437–472, 1998.

33. Z. Lao, A. Thiede, U. Nowotny, H. Lienhart, V. Hurm, M. Schlechtweg, J. Hornung, W. Bronner, K. Kohler, A. Hulsmann, B. Raynor, and T. Jakobus, "40-Gb/s high-power modulator driver IC for lightwave communication systems," *IEEE J. Solid State Circuits*, 33, 1520–1526, 1998.

34. H. Shigematsu, N. Yoshida, M. Sato, N. Hara, T. Hirose, and Y. Watanabe, "45-GHz distributed amplifier with a linear 6-Vp-p output for a 40-Gb/s LiNbO₃ modulator driver circuit," *Tech. Dig. IEEE GaAs IC Symp.*, 137–140, 2001.

35. L.D. Nguyen, L.E. Larson, and U.K. Mishra, "Ultra-high-speed modulation-doped field-effect transistors: a tutorial review," *Proc. IEEE*, 80, 494–518, 1992.

36. T. Enoki, H. Ito, K. Ikuta, Y. Umeda, and Y. Ishii, "0.1-μm InAlAs/InGaAs HEMTs with an InP-recess-etch stopper grown by MOCVD," *Microwave and Optical Tech. Lett.*, 11, 135–139, 1996.

37. T. Suemitsu, T. Ishii, and Y. Ishii, "Gate and recess engineering for ultrahigh-speed InP-based HEMTs," *Dig. Topical Workshop on Heterostructure Microelectronics*, 2–3, 2000.

38. Y. Yamashita, A. Endoh, K. Shinohara, M. Higashiwaki, K. Hikosaka, T. Mimura, S. Hiyamizu, and T. Matsui, "Ultra-short 25-nm-gate lattice-matched InAlAs/InGaAs HEMTs within the range of 400 GHz cutoff frequency," *IEEE Electron Device Lett.*, 22, 367–369, 2001.

39. S. Kimura, Y. Imai, Y. Umeda, and T. Enoki, "0–90 GHz InAlAs/InGaAs/InP HEMT distributed amplifier IC, *IEE Electron. Lett.*, 31, 1430–1431, 1995.

40. K. Murata, T. Otsuji, E. Sano, S. Kimura, and Y. Yamane, "70-Gbit/s multiplexer and 50-Gbit/s decision IC modules using InAlAs/InGaAs/InP HEMTs," *IEICE Trans. Electron.*, E83-C, 1166–1169, 2000.

41. K. Sano, K. Murata, and Y. Yamane, "50-Gbit/s demultiplexer IC module using InAlAs/InGaAs/InP HEMTs," *IEICE Trans. Electron.*, E83-C, 1786–1788, 2000.

42. K. Murata, K. Sano, E. Sano, S. Sugitani, and T. Enoki, "Fully monolithic integrated 43 Gbit/s clock and data recovery circuit in InPHEMT technology," *IEE Electron. Lett.*, 37, 1235–1237, 2001.

43. Y. Nakasha, T. Suzuki, H. Kano, A. Ohya, K. Sawada, K. Makiyama, T. Takahashi, M. Nishi, T. Hirose, and Y. Watanabe, "A 43Gb/s full-rate-clock 4:1 multiplexer in InP-based HEMT technology," *Dig. Tech. Papers IEEE Int. Solid-State Circuits Conf. (ISSCC)*, paper No. 11.6, 2002.

44. T. Suzuki, Y. Nakasha, T. Sakoda, K. Sawada, T. Takahashi, K. Makiyama, T. Hirose, and M. Takigawa, "A 100-Gbit/s 2:1 Multiplexer in InP HEMT Technology," *Dig. IEEE MTT-S Int. Microwave Symp.*, WE5D-1, 2003.

45. Y. Kuriyama, T. Sugiyama, S. Hongo, J. Akagi, K. Tsuda, N. Iizuka, and M. Obara, "A 40GHz D-type flip-flop using AlGaAs/GaAs HBTs," *Tech. Dig. IEEE GaAs IC Symp.*, 189–192, 1994.

46. T. Ishibashi and Y. Yamauchi, "A novel AlGaAs/GaAs HBT structure for near-ballistic collection," *Dig. 45th Annual Device Res. Conf.*, p. IVA-6, 1987.

47. T. Ishibashi, Y. Yamauchi, E. Sano, H Nakajima, and Y. Matsuoka, "Ballistic collection transistors and their applications," *Int. J. High Speed Electron. Sys.*, 5, 349–379, 1994.

48. K. Runge, P.J. Zampardi, R.L. Pierson, R. Yu, P.B. Thomas, S.M. Beccue, and K.C. Wang, "AlGaAs/GaAs HBT circuits for optical TDM communications," *Int. J. High Speed Electron. Sys.*, 9, 473–503, 1999.

49. R. Ohhira, Y. Amamiya, T. Niwa, N. Nagano, T. Takeuchi, C. Kurioka, T. Chuzenji, and K. Fukuchi, "AlGaAs/InGaAs HBT IC modules for 40-Gb/s optical receiver," *IEICE Trans. Electron.*, E82-C, 448–455, 1999.

50. H. Kroemer, "Heterostructure bipolar-transistors—what should we build," *J. Vacuum Sci. Tech.*, B 1, 126–130, 1983.

51. T.S. Low, C.P. Hutchinson, P.C. Canfield, T.S. Shirley, R.E. Yeats, J.S.C. Chang, G.K. Essilfie, M.K. Culver, W.C. Whiteley, D.C. D'Avanzo, N. Pan, J. Elliot, and C. Lutz, "Migration from an AlGaAs to an InGaP emitter HBT IC process for improved reliability," *Tech. Dig. IEEE GaAs IC Symp.*, 153–156, 1998.

52. K. Mochizuki, T. Oka, K. Ouchi, and T. Tanoue, "Reliability investigation of heavily C-doped InGaP/GaAs HBTs operated under a very high current-density condition," *Solid-State Electron.*, 43, 1425–1428, 1999.

53. T. Oka, K. Hirata, H. Suzuki, K. Ouchi, H. Uchiyama, T. Taniguchi, K. Mochizuki, and T. Nakamura, "High-speed small-scale InGaP/GaAs HBT technology and its application to integrated circuits," *IEEE Trans. Electron Devices*, 48, 2625–2630, 2001.

54. K. Uchida, H. Matsuura, T. Yakihara, S. Kobayashi, S. Oka, T. Fujita, and A. Miura, "A series of InGaP/InGaAs HBT oscillators up to D-band," *IEEE Trans. Microwave Theory Tech.*, 49, 858–865, 2001.

55. M. Ida, K. Kurishima, and N. Watanabe, "Over 300 GHz f_T and f_{max} InP/InGaAs Double Heterojunction Bipolar Transistors With a Thin Pseudomorphic Base," *IEEE Electron Devices. Lett.*, 23, 694–696, 2002.

56. S. Yamahata, K. Kurishima, H. Ito, and Y. Matsuoka, "Over-220-GHz-f_T- and-f_{max} InP/InGaAs double-heterojunction bipolar transistors with a new hexagonal-shaped emitter," *Tech. Dig. IEEE GaAs IC Symp.*, 163–166, 1995.

57. S. Yamahata, H. Nakajima, M. Ida, H. Niiyama, N. Watanabe, E. Sano, and Y. Ishii, "Reliable carbon-doped InP/InGaAs HBTs technology for low-power 40-GHz static frequency divider," *Proc. Int. Conf. Solid State Devices and Materials*, 570–571, 1999.

58. E. Sano, H. Nakajima, N. Watanabe, and S. Yamahata, "40 Gbit/s decision IC using InP/InGaAs composite-collector heterojunction bipolar transistors," *IEE Electron. Lett.*, 35, 1194–1195, 1999.

59. E. Sano, H. Nakajima, N. Watanabe, S. Yamahata, and Y. Ishii, "40 Gbit/s 1:4 demultiplexer IC using InP-based heterojunction bipolar transistors," *IEE Electron. Lett.*, 35, 2116–2117, 1999.

60. M.J.W. Rodwell, Miguel Urteaga, D. Mensa, Q. Lee, J. Guthrie, Y. Betser, S.C. Martin, R.P. Smith, S. Jaganathan, T. Mathew, P. Krishnan, S. Long, R. Pullela, B. Agarwal, U. Bhattacharya, L. Samoska, D. Scott, and M. Dahlstrom, "Submicron Scaling of HBTs," *IEEE Trans. Electron Devices*, 48, 2606–2624, 2001.

61. T. Mathew, H.J. Kim, S. Jaganathan, D. Scott, S. Krishnan, Y. Wei, M. Urteaga, M.J.W. Rodwell, and S. Long, "75 GHz ECL Static Frequency Divider Using InAlAs/InGaAs HBTs," *IEE Electron. Lett.*, 37, 667–668, 2001.

62. C.H. Fields, M. Sokolich, S. Thomas, K. Elliot, and J. Jensen, "Progress toward 100 GHz logic in InP HBT IC Technology," *Int. J. High Speed Electron. Sys.*, 11, 217–243, 2001.

63. M. Sokolich, C. Fields, B. Shi, Y.K. Brown, M. Montes, R. Martinez, A.R. Kramer, S. Thomas III, and M. Madhav, "A low power 72.8 GHz static frequency divider implemented in AlInAs/InGaAs HBT IC technology," *Tech. Dig. IEEE GaAs IC Symp.*, E.11, 2000.

64. J.P. Mattia, R. Pullela, G. Georgieu, Y. Baeyens, H.S. Tsai, Y.K. Chen, C. Dorschky, T. Winkler Von Mohrenfels, M. Reinhold, C. Groepper, M. Sokolich, L. Ngyuen, and W. Stanchina, "High-speed multiplexers: a 50Gb/s 4:1 MUX in InP HBT technology," *Tech. Dig. IEEE GaAs IC Symp.*, 189–192, 1999.

65. J. Godin, P. Andre, J.L. Benchimol, P. Berdaguer, S. Blayac, J.R. Burie, P. Desrousseaux, A.M. Duchenois, N. Kauffmann, A. Konczykowska, and M. Riet, "40 Gbits optical communications: InP DHBT technology, circuits and system experiments," *Tech. Dig. IEEE GaAs IC Symp.*, 185–188, 1999.

66. A. Kasbari, P. Andre, J. Godin, and A. Konczykowska, "40 Gbit/s master-slave D-type flip-flop in InP DHBT technology," *IEE Electron. Lett.*, 38, 330–331, 2002.

67. M. Dahlström1, M. Urteaga, S. Krishnan, N. Parthasarathy, and M.J.W. Rodwell, "Ultra-Wideband DHBTs using a Graded Carbon-Doped InGaAs Base," postdeadline paper, Indium Phosphide and Related Materials Conf. (IPRM), 2002.

68. E. Sano, K. Sano, T. Otsuji, K. Kurishima, and S. Yamahata, "Ultra-high speed low-power monolithic photoreceiver using InP/InGaAs double-heterojunction bipolar transistors," *IEE Electron. Lett.*, 33, 1047–1048, 1997.
69. V. Hurm, W. Benz, W. Bronner, A. Hulsmann, T. Jakobus, K. Kohler, A. Leven, M. Ludwig, B. Raynor, J. Rosenzweig, M. Schlechtweg, and A. Thiede, "40 Gbit/s 1.55 μm pin-HEMT photoreceiver monolithically integrated on 3in GaAs substrate," *IEE Electron. Lett.*, 34, 2060–2062, 1998.
70. D. Huber, M. Bitter, T. Morf, C. Bergamashi, H. Melchior, and H. Jackel, "46 GHz bandwidth monolithic InP/InGaAs pin/SHBT photoreceiver," *IEE Electron. Lett.*, 35, 40–41, 1999.
71. K. Takahata, Y. Muramoto, H. Fukano, K. Kato, A. Kozen, S. Kimura, Y. Imai, Y. Miyamoto, O. Nakajima, and Y. Matsuoka, "Ultrafast monolithic receiver OEIC composed of multimode waveguide p-i-n photodiode and HEMT distributed amplifier," *IEEE J. Selected Top. Quantum Electron.*, 6, 31–37, 2000.
72. T. Ishibashi, N. Shimizu, S. Kodama, H. Ito, T. Nagatsuma, and T. Furuta, "Uni-traveling-carrier photodiodes," *Tech. Dig. OSA Int. Topical Meeting on Ultrafast Electron. and Optoelectron.*, 166–168, 1997.
73. K. Murata, H. Kitabayashi, S. Shimizu, S. Kimura, T. Furuta, N. Watanabe, and E. Sano, "A 40-Gbit/s monolithic digital OEIC module composed of uni-traveling-carrier photodiode and InP HEMT decision circuit," *Dig. IEEE MTT-S Int. Microwave Symp.*, 345–348, 2000.
74. T. Otsuji, M. Yoneyama, K. Murata, and E. Sano, "A super-dynamic flip-flop circuit for broadband applications up to 24 Gb/s utilizing production-level GaAs MESFETs," *IEEE J. Solid State Circuits*, 32, 1357–1362, 1997.
75. K. Murata, T. Otsuji, M. Yoneyama, and M. Tokumitsu, "A 40-Gbit/s super-dynamic decision IC fabricated with 0.15-μm GaAs MESFETs," *IEEE J. Solid-State Circuits*, 33, 1527–1535, 1998.
76. T. Otsuji, K. Murata, T. Enoki, and Y. Umeda, "An 80-Gbit/s multiplexer IC using InAlAs/InGaAs/InP HEMTs" *IEEE J. Solid-State Circuits*, 33, 1321–1327, 1998.
77. K. Murata, T. Otsuji, M. Ohhata, M. Togashi, E. Sano, and M. Suzuki, "A novel high-speed latching operation flip-flop (HLO-FF) circuit and its application to a 19 Gb/s decision circuit using 0.2 μm GaAs MESFET," *IEEE J. Solid State Circuits*, 30, 1101–1108, 1995.
78. K. Ishii, H. Ichino, M. Togashi, Y. Kobayashi, and C. Yamaguchi, "Very-high-speed Si bipolar static frequency-divider with new T-type flip-flops," *IEEE J. Solid-State Circuits*, 30, 19–24, 1995.
79. B. Tang, J. Northoff, A. Gutierres-Aitken, E. Kaneshiro, and P. Chin, "InP DHBT 68 GHz frequency divider," *Tech. Dig. IEEE GaAs IC Symp.*, 193–196, 1999.
80. T. Suzuki, H. Kano, Y. Nakasha, T. Takahashi, K. Imanishi, H. Ohnishi, and Y. Watanabe, "40-Gbit/s D-type flip-flop and multiplexer circuits using InP HEMT," *Dig. IEEE MTT-S Int. Microwave Symp.*, TUIF-50–1–4, 2001.
81. M. Sokolich, G. Raghavan, D.A. Hitko, and Y. Brown, "InP HBT Production Technology for 100 Gbps Lightwave Communications," *GaAs ManTech Digest of Papers*, 99, 169–172, 1999.
82. S. Subbanna, J. Johnson, G. Freeman, R. Volant, R. Groves, D. Herman, and B. Meyerson, "Prospects of silicon-germanium-based technology for very high-speed circuits," *Dig. IEEE MTT-S Int. Microwave Symp.*, 1, 361–364, 2000.
83. T. Enoki, H. Yokoyama, Y. Umeda, and T. Otsuji, "Ultrahigh-speed integrated circuits using InP-based HEMTs," *Jpn. J. Appl. Phys.*, 37, 1359–1346, 1998.

84. M. Matloubian, D. Docter, C. Nguyen, T. Liu, S. Bui, and C. Ngo, "$Ga_{0.47}In_{0.53}As$/InP HEMTs with novel $GaP_{0.35}Pb_{0.65}$ Schottky layers grown by MOVPE," *Dig. 56th Annu. Dev. Res. Conf.*, 32–33, 1998.

85. C.R. Bolognesi, M. Matine, M.W. Dvorak, P. Yeo, X.G. Xu, and S.P. Watkins, "InP/GaAsSb/InP Double HBTs: a new alternative for InP-based DHBTs," *IEEE Trans. Electron Devices*, 48, 2631–2639, 2001.

86. C. Marrian, "DARPA Antimonide Based Compound Semiconductors Program," Third Workshop on the Fabrication, Characterization, and Applications of 6.1 Å III-V Semiconductors, 2001.

87. K. Maezawa and T. Mizutani, "A new resonant tunneling logic gate employing monostable-bistable transition," *Jap. J. Appl. Phys.*, 37, 1359–1364, 1993.

88. K. Sano, K. Murata, T. Otsuji, T. Akeyoshi, N. Shimizu, M. Yamamoto, T. Ishibashi, and E. Sano, "Ultra-fast optoelectronic decision circuit using resonant tunneling diodes and a uni-traveling-carrier photodiode," *IEICE Trans. Electron.*, E82-C, 1638–1646, 1999.

89. K. Sano, K. Murata, T. Otsuji, T. Akeyoshi, M. Shimizu, and E. Sano, "An 80-Gb/s optoelectronic delayed flip-flop IC using resonant tunneling diodes and uni-traveling-carrier photodiode," *IEEE J. Solid State Circuits*, 36, 281–289, 2001.

90. K. Murata, K. Sano, T. Akeyoshi, N. Shimizu, E. Sano, M. Yamamoto, and T. Ishibashi, "An optoelectronic clock recovery circuit using a resonant tunneling diode and a uni-traveling-carrier photodiode," *IEICE Trans. Electron.*, E82-C, 1494–1501, 1999.

91. K. Narahara, T. Otsuji, N. Shimizu, and E. Sano, "An optically-controlled phase shifter for traveling-wave retiming circuit," in *OSA Trends in Optics and Photonics*, Vol. 28, Ultrafast Electronics and Optoelectronics, J.E. Bowers and W.H. Knox, Eds., Optical Society of America, Washington, DC, 211–218, 1999.

92. T. Otsuji, "Present and future of high-speed compound semiconductor IC'S," *Int. J. High Speed Electron. Sys.*, 13, 1–25, 2003.

93. K. Narahara and T. Otsuji, "A traveling-wave time-division demultiplexer," *Jpn. J. Appl. Phys.*, 38, 4021–4026, 1999.

8 High-Speed All-Optical Technologies for Photonics

Masatoshi Saruwatari

CONTENTS

8.1 INTRODUCTION

To accommodate the coming broadband era, very high-speed photonic technologies must be developed not only for transmission lines, but also for transmission nodes. The goal is to handle signal rates of more than several terabit/s so that vast amounts of information, including data and moving pictures, can be provided to many subscribers through optical fiber cables. To this end, novel all-optical signal processing technologies capable of superseding conventional electron-based technologies have been extensively studied. The key technologies include ultrashort optical pulse generation and modulation, all-optical multiplexing, all-optical demultiplexing, optical timing extraction, optical signal monitoring, optical linear or nonlinear (soliton) transmission, all-optical repeating, all-optical regenerating (2R or 3R), and so on. This chapter describes very high-speed photonic technologies required for multi-terabit/s optical fiber transmission systems employing optical time-division multiplexing (OTDM) and time- and wavelength-division multiplexing (OTDM/WDM) techniques, focusing on ultrafast pulse generation, all-optical multiplexing, all-optical demultiplexing, optical timing extraction, and very high-speed optical waveform measurement based on optical sampling.

8.2 KEY TECHNOLOGIES FOR VERY HIGH-SPEED TRANSMISSION

8.2.1 ULTRASHORT OPTICAL PULSE GENERATION

To achieve very high-speed OTDM transmission systems, it is essential for optical sources to generate picosecond transform-limited (TL) chirpless pulses at repetition rates ranging from 10 to 40 GHz. In addition, the repetition rate should be tunable and controllable to permit synchronization with other signals, and oscillation wavelength should be tunable to optimize picosecond pulse transmission. Table 8.1 compares various optical pulse generation techniques that have been applied to high-speed (>10 Gbit/s) optical transmission experiments. These techniques include gain-switching of distributed-feedback laser-diodes (DFB-LDs), gating of continuous-wave (CW) light with an electroabsorption (EA) modulator, mode-locking of laser diodes (LDs), harmonic mode-locking of erbium- (Er-) doped fiber (EDF) lasers, and supercontinuum (SC) generation by a dispersion-decreasing fiber pumped with an intense picosecond pulse. In the following section, these techniques will be described.

TABLE 8.1
High-Speed Optical Pulse Generation Techniques

Method	Repetition Frequency [limitation]	$\Delta\tau$ (ps)	$\Delta\tau \cdot \Delta\nu$	Comment
Gain-switching of DFB-LD	arbitrary [RC constant]	~ 20	> TL	down-chirping
		(6)	(~ TL)	(with chirp compensation: CC)
		(0.8)	(~ TL)	(with CC + soliton adiabatic compression)
CW + EA-modulator	arbitrary [modulator]	~ 15	~ TL	relatively large pulse width
		(2.5)	(~ TL)	(with adiabatic soliton compression)
Mode-locking of LD	fixed	~ 10	> TL	conventional type
		< 1	~ TL	CPM-type or EA modelocker
Harmonic mode-locking of EDF laser	tunable [mode locker]	1–3	TL	wavelength-tunable, – 20 nm
SC generation in DDF	tunable [pump frequency]	< 1	~ TL	wavelength-tunable, >200 nm

Note: EA = electroabsorption; TL = transform-limited; SC = supercontinuum; DDF = dispersion-decreasing fiber.

8.2.1.1 Gain-Switching of DFB-LDs

When a DFB-LD is driven by a sinusoidal current of several GHz, 25–30 ps optical pulses with down (red-shift) chirping can be generated. When the gain-switched pulses transit through a dispersion-shifted fiber with normal dispersion of 45 ps/nm, the down chirping can be compensated for, resulting in nearly TL-compressed 5 to 7-ps pulses [1]. Because the repetition rate can be easily tuned to an arbitrary frequency between 1–20 GHz, gain-switching has been applied for early high-speed experiments, including soliton transmission, all-optical demultiplexing, optical timing extraction and optical sampling. To further reduce gain-switched pulses to < 3 ps duration, however, nonlinear pulse-compression techniques such as an adiabatic soliton compression technique should be incorporated together with the linear chirp compensation.

8.2.1.2 Gating of CW Light with an EA Modulator

The second pulse-generation technique is gating of CW light with sinusoidally driven EA modulators [2,3]. This is based on the nonlinear transmittance of InGaAsP-EA modulators with respect to applied voltage, and nearly transform-limited 20-ps soliton pulses have been obtained at 10 Gbit/s. Generated soliton pulses were transmitted over a recirculating distance greater than 6000 km. However, this technique cannot be applied to higher-speed TDM systems such as 100 Gbit/s due to its relatively large pulse width (> 10 ps). To improve this, a pulse-compression technique using a dispersion-decreasing fiber as a soliton adiabatic compressor has been demonstrated to obtain 2.5 ps TL pulses [3].

8.2.1.3 Mode-Locking of LDs

Mode-locking of LDs [4–10] is an effective way of producing high-repetition-rate subpicosecond optical pulses. In particular, a novel type of LD having a saturable absorber at the midpoint of the cavity is intriguing because of its colliding-pulse mode-locking (CPM) operation [4]. This device outputs ultrahigh-speed, purely TL pulses with less than 1-ps duration at 40 GHz and 350 GHz repetition frequency through active and passive CPM operation, respectively. At that time, there was no easy way to utilize this extremely high operation speed. In order to mode-lock at moderate speed, a long-cavity LD integrated with a passive waveguide and a Bragg reflector was developed [5]. The cavity length was monolithically extended to 5.5 mm, leading to 8-GHz mode-locking operation. This was used for 8-Gbit/s 4000-km soliton transmission experiments. The main drawback of this method is that there is no tunability of repetition frequency and relatively larger spectral width as compared with the TL condition.

To improve the operation characteristics, various mode-locked LDs have been studied, including pulse-width tunable subpicosecond pulse generation from an active mode-locked monolithic multiquantum well (MQW), LDs integrated with MQW-EA modulators [6], and repetition rate tunable lasers using passive mode-locking of micromechanically-tunable LDs [7] or active mode-locking of external cavity LDs [8]. Also reported are femtosecond optical pulse generation from active mode-locked LDs by compensating for higher-order chirping using fibers with different dispersions [9] and very high-speed optical pulse generation at 100 GHz with low jitter from active mode-locked LDs integrated with EA modulators [10]. At present, mode-locked LDs are expected to be high-speed optical signal sources applied for the next 40Gbit/s systems and supercontinuum pump sources due to its preferable features such as compactness, high stability, and low cost.

8.2.1.4 Harmonic Mode-Locking of EDF Lasers

Harmonic mode-locking of EDF lasers [11–17] is promising because it can generate a purely TL pulse train in the 10-GHz region without requiring any chirp compensation or pulse compression. However, since EDF lasers are normally tens of meters long, they have to be mode-locked at very high harmonics of the cavity mode spacing, for example, harmonic orders on the order of 1000. This leads to the issue of how to stabilize such laser systems in which the competition between many sets of supermodes is apt to occur. Moreover, mode competition is enhanced by fluctuations in cavity lengths and polarization states. To stabilize mode-locked EDF lasers, several techniques have been developed. The first method is inserting a high-finesse Fabry-Perot etalon in the cavity to eliminate all the unwanted laser cavity modes [12]. However, this straightforward method requires the strict condition that the free-spectral range of the etalon should coincide with the mode-locking frequency. The second method is dithering the cavity length at a kHz rate [13] to wash out spatial hole burning. Although this makes one set of supermodes dominant, polarization fluctuations still remain, resulting in unstable operation. The third method uses a polarization-maintaining cavity to eliminate the mode-competition generated by polarization fluctuations [14–17]. Figure 8.1 depicts a wavelength-tunable mode-locked Er fiber ring laser based on

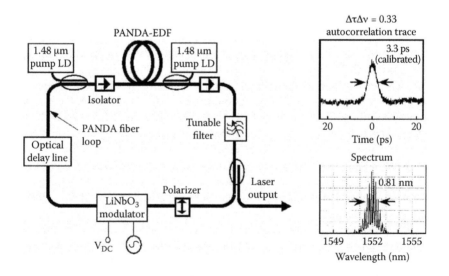

FIGURE 8.1 Configuration of harmonic mode-locked Er fiber laser.

this principle. In this case, all the fibers, including the Er-doped fiber, are polarization-maintaining (PM) single-mode fibers, called PANDA fibers, and a pigtailed polarizer is inserted to ensure single-polarization oscillation. A tunable optical filter and an optical delay line are inserted into the cavity to obtain wavelength tunability. With this configuration, 3.0- to 3.5-ps TL pulses having a time-bandwidth product of 0.33 and corresponding to $\text{sech}^2(t)$ waveform were obtained at up to 20 GHz. Stable oscillation characteristics without any pulse-to-pulse fluctuations were also verified by bit-error-rate (BER) measurements using external modulation [15]. Incidentally, wavelength tunable operation over 7 nm was applied to optimize the optical demultiplexer operation and OTDM transmission experiments. Since EDF lasers have relatively long cavities, temperature change brings about considerable cavity length variations, resulting in unstable mode-locking. To stabilize the mode-locking, the cavity length of EDF lasers is actively controlled by an optical delay line in the feedback loop [16]. Since the onset of relaxation oscillation was found to be a good measure of the detuning, its RF power was used for the feedback signal. With this method, a very stable error-free operation over 10 hours has been confirmed [16]. More recently, a stable fixed-frequency operation of mode-locked EDF lasers has been reported using the improved regenerative mode-locking method that controls the cavity length by a phase-locked loop (PLL) using the extracted clock from mode-locked pulses [17].

8.2.1.5 Supercontinuum (SC) Pulse Generation

When an intense optical pulse transits a low-dispersion silica fiber, a supercontinuum (SC) pulse, whose spectrum is spread over a wide spectral range, is induced by the combining effects of various optical nonlinear processes [18–22]. The first SC pulse generation applicable to optical transmissions was achieved by using 1542-nm, 3-ps TL pulses from a 6.3-GHz mode-locked EDF laser, as shown in Figure 8.2. With a 4-W peak, a "white" pulse with a 200-nm-wide spectrum ranging from 1450 nm to 1650 nm was generated from a 3-km dispersion-shifted fiber. The SC spectrum does not only exhibit 200-nm continuous, flat-top broadening, but also has some degree of coherency. Recently, it has been clarified that the desired flatly broadened SC spectra can be generated from a dispersion-decreasing fiber with a convex profile as a function of wavelength, having two zero-dispersion wavelengths at the input [20]. With the use of these characteristics, penalty-free 3-ps pulse generation for WDM systems has been demonstrated, where many WDM channels can be produced at arbitrary wavelengths by merely filtering out the desired wavelengths [19]. Furthermore, by using a 5-nm bandpass filter and chirp compensation using a 1.3-μm zero-dispersion fiber, nearly TL pulses with shorter than 0.2 ps can be obtained within a wide spectral range. By filtering a super-broadened SC spectrum with an arrayed-waveguide grating (AWG)-WDM demultiplexer, a multiwavelength pico-second pulse source has been constructed for dense WDM/TDM systems. For example, using an SC generator and a 16×16 AWG-WDM demultiplexer with 100-GHz frequency spacing, 16 different-wavelength 6.3-GHz optical pulses can be simultaneously generated from a single SC pulse source. The output WDM pulses feature a very low jitter (< 0.3 ps) and high frequency stability (1.26 GHz/°C) determined by the pump source and AWG characteristics, respectively.

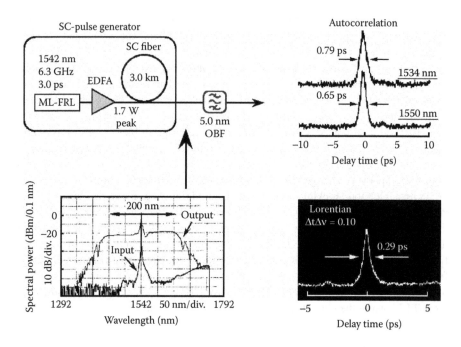

FIGURE 8.2 Configuration of supercontinuum (SC) pulse generation. Output spectra and pulse waveform are shown.

More recently, it has been found that an SC pulse source can generate stable CW multiple carriers with an equal frequency spacing by merely slicing each spectrum of longitudinal modes with an AWG-WDM demultiplexer [21,22]. The basic configuration of supercontinuum based multicarrier sources is depicted in Figure 8.3. Although conventional mode-locked lasers have a limited number of multicarriers determined by their original pulse width, the SC source can generate a great number of carriers due to its super-broadened coherent spectrum. For example, over 1000 channel optical frequency chain generation with 12.5-GHz channel spacing [21], and also 50-GHz spaced optical carrier generation matched to the optical ITU grid seamlessly over S-, C-, and L-bands [22], have been successfully demonstrated using a single SC source.

8.2.2 ALL-OPTICAL MULTIPLEXER/DEMULTIPLEXER

Ultrahigh-speed time-division multiplexing/demultiplexing (MUX/DEMUX), where the channels of the lower digital hierarchy signals are multiplexed to those of the higher hierarchy signals (MUX) and vice versa (DEMUX) in the time domain, is one of the key functions necessary to achieve terabit/s optical transmission. The operation speed of conventional MUX/DEMUX based on electronics is currently limited to around 40 Gbit/s. To circumvent such electronics bottlenecks, all-optical MUX/DEMUXs utilizing ultrafast third-order optical nonlinearities (e.g., nonlinear refractive index n_2) are promising.

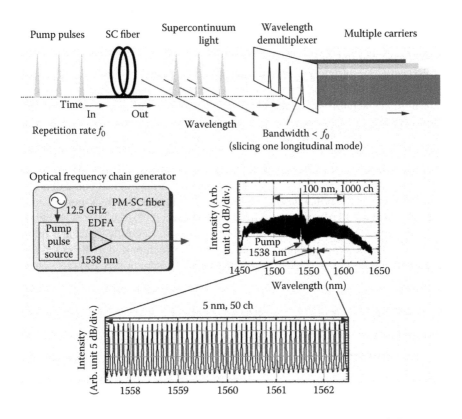

FIGURE 8.3 Multicarrier generation from a single SC source. (After Takara [21].)

8.2.2.1 All-Optical Multiplexer (MUX)

To realize a very high-speed OTDM system, a stable optical time-domain multiplexer (MUX) is needed. Although much attention has been paid to all-optical DEMUX, a compact MUX using a planar lightwave circuits called PLC was fabricated. The PLC-MUX is based on the multi-stage interferometer-type multiplexer in which 2-by-2 couplers and optical delay line waveguides are formed on a silicon substrate by a chemical vapor deposition method. With this configuration, the original 6.3-Gbit/s signal was stably multiplexed to a 100-Gbit/s OTDM test signal. Another PLC-MUX was also developed, consisting of delay lines and a couple of multimode waveguides as branching or combining devices. Since both devices are connected with an integer number of delay lines with the same length difference, an arbitrary number of channels can be multiplexed with this configuration. Although these PLC-MUXs are not real MUXs combining different channels, they are very efficient for generating stable high-speed test signals far above the available modulation speed more than 100 Gbit/s.

All-optical modulation applied for MUX operation was reported using four-wave mixing (FWM) in a traveling-wave semiconductor optical amplifier (TW-SOA) [23,24]. A higher repetition rate (100 GHz) optical pulse train is all-optically modulated by full sets of lower-speed (10 Gbit/s) optical signals to make a fully multiplexed

OTDM signal. In [24], two-channel error-free multiplexing of 10 Gbit/s individual signals on a 100-GHz optical clock was demonstrated using FWM in two sets of SOAs connected in series.

Recently, an intriguing and simple method has been demonstrated to generate a multiplied optical clock from an original optical clock [25]. This consists of an ultrashort optical pulse source and a highly dispersive fiber to cause a linearly chirped pulse broadening. If the chirped pulse broadening exceeds the pulse interval, the chirped pulses partially overlap each other. When the frequency difference Δv between the overlapped pulses at the same temporal position satisfies $\Delta v = Mf_0$, where f_0 is the original repetition rate and M is an integer, a harmonic pulse train with a repetition frequency M times higher than that of the original one will appear. Figure 8.4 shows the principle and experimental results of this clock-multiplication method. Using a 10-GHz pulse train from a mode-locked EDF laser or SC generator, a 50-GHz or 400-GHz optical pulse train has been successfully generated by this method. Since the generated pulse train is perfectly synchronized to the initial train, an all-optical 100-Gbit/s MUX experiment, where the multiplied 100-GHz optical

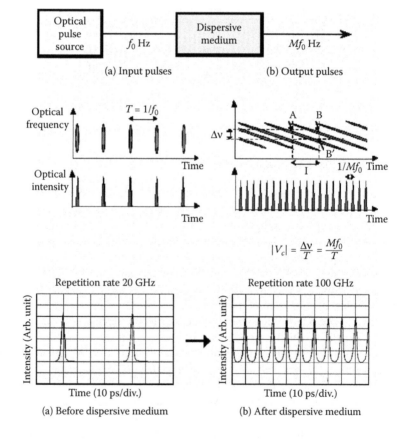

FIGURE 8.4 Principle of optical clock multiplication using dispersive media.

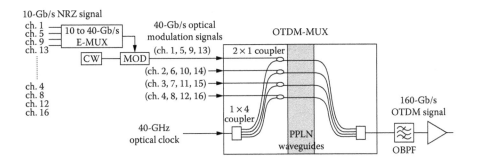

FIGURE 8.5 Fundamental configuration of integrated OTDM-MUX.

clock is all-optically modulated by three 10-Gbit/s optical channels with a nonlinear optical loop mirror (NOLM) switch, has been successfully demonstrated [26].

Recently, 120- to 160-Gbit/s OTDM-MUX prototypes have been developed aiming at post–40-Gbit/s systems. They include a 120-Gbit/s (20 Gbit/s × 4 ch + 40 Gbit/s) OTDM-MUX prototype using LN modulators with optical fiber delay lines [27], and a 160-Gbit/s (40 Gbit/s × 4 ch) OTDM-MUX prototype using an EA-modulator based-OTDM module and a 40-GHz external-cavity mode-locked LD [28], each of which has been applied for a single-channel transmission experiment of 120 Gbit/s (160 km) or 160 Gbit/s (300 km).

More recently, a full 160-Gbit/s (40 Gbit/s × 4 ch) OTDM-MUX has been developed using a hybrid integrated PLC circuit. This circuit has an arrayed wavelength converter based on FWM in quasi–phase-matched LiNbO$_3$ (QPM-LN) waveguides for all-optical modulation, together with two built-in couplers with delay lines for 1 × 4 splitting and 4 × 1 combining, as shown in Figure 8.5 [29, 30]. The full MUX circuit has been applied for the first 160-Gbit/s OTDM transmission having full all-channel modulation together with full all-optical demultiplexing [30], as described later.

8.2.2.2 All-Optical Demultiplexer (DEMUX)

All-optical demultiplexers (DEMUXs) are key devices for developing ultrafast terabit/s OTDM transmission systems. To apply the DEMUXs to real transmission systems, the following requirements should be satisfied: fast and stable bit-error-free operation, low control power capable of LD or EDF laser pumping, polarization-insensitive (PI) function, and multiple output (full DEMUX) operation. Here, various all-optical demultiplexing techniques based on third-order optical nonlinearity will be reviewed, taking into account these requirements.

Table 8.2 summarizes the various all-optical DEMUX techniques reported so far. These are the optical Kerr switch [31–33], four-wave mixing (FWM) using a fiber [34–37] or an SOA [38–41], cross-phase modulation (XPM)-based switching using a fiber [42–45] or another XPM-based switching using SOAs called differential-phase modulation (DPM) [60–62], and various fiber loop based switches, called NOLM [46–54], SLALOM (semiconductor laser amplifier in a loop mirror configuration) [55,56], or TOAD (terahertz optical asymmetric demultiplexer) [57–59]. In most cases, LD pumping has been realized with the aid of Er-doped fiber amplifiers

TABLE 8.2
All-Optical Demultiplexing Techniques

Concept		Bit rate (Gbit/s)	BER	Author	Year	Ref.
Optical Kerr Switch		2/60	—	Morioka (NTT)	1987/92	31, 32
		40	—	Patrick (BT)	1993	33
FWM	Fiber	16	free	Andrekson (AT&T)	1991	34
Switch		100-PI	free	Morioka (NTT)	1994	35
		100 (10 G × 4 ch)	free	Morioka (NTT)	1994	36
		500	free	Morioka (NTT)	1996	37
	SOA	20	free	Ludwig (HHI)	1993	38
		100	free	Kawanishi (NTT)	1994	39
		200-PI	free	Morioka (NTT)	1996	40
		160 (20 G × 8 ch)	free	Shake (NTT)	2002	41
XPM	Fiber	60-PI	—	Morioka (NTT)	1992	42
Switch		40	—	Patrick (BT)	1993	43
		100 (10 G × 6 ch)	free	Uchiyama (NTT)	1997	44
		100 (10 G × 10 ch)	free	Uchiyama (NTT)	2001	45
	SOA(DPM)	168/336	free	Nakamura (NEC)	2001/02	60, 61
Loop	NOLM	1	—	Blow (BT)	1990	46
Mirror		5	—	Jinno (NTT)	1990	47
		40	—	Takada (NTT)	1991	48
		64	free	Andrekson (AT&T)	1992	49
		40	free	Patrick (BT)	1993	50
		20-PI	—	Bülow (Alcatel SEL)	1993	51
		32/100-PI	free	Uchiyama (NTT)	1993/94	52
		100 (5 ch)	free	Uchiyama (NTT)	1996	53
		640	free	Yamamoto (NTT)	1998	54
	SLALOM	9	—	Eiselt (HHI)	1993	55
		80	free	Diez (HHI)	1998	56
	TOAD	50/250	—	Sokoloff (Princeton Univ.)	1993/94	57
		40	free	Ellis (BT)	1993	58
		160	free	Suzuki (NTT)	1994	59

Note: BER = bit error rate; free = bit-error-free operation; ch = number of channels of multiple-output operation; PI: polarization-independent operation.

(EDFA). Polarization independence or multiple-output operation has also been demonstrated in several ways. The maximum operation speed reported so far is 500 Gbit/s using FWM in a fiber and 640 Gbit/s using NOLM switching. In the following section, some of the DEMUX techniques will be described in more detail.

Optical Kerr Switch — An optical Kerr switch is based on the 90°-polarization rotation due to the induced birefringence caused by intense optical pulses. The first high-speed DEMUX operation at 2 Gbit/s was demonstrated with the optical Kerr switch using a pair of identical polarization-maintaining (PM) fibers, so-called PANDA fibers, being cross-spliced to compensate for natural birefringence [31]. Moreover, a reflective Kerr DEMUX operation at 60 Gbit/s was achieved by a complete birefringence-compensation technique using a "polarization-rotating mirror" that reflected the input signal and pump pulses with their polarization states flipped by 90° [32]. However, in principle, it is difficult for the Kerr switch to operate with polarization independence.

FWM Switch in Fibers — The second intriguing method is the use of the four-wave mixing (FWM) process that occurs in silica fibers or semiconductor optical amplifiers (SOAs). The general principle of FWM-based demultiplexing is as follows: When pump pulses at wavelength λ_p propagate synchronously with original pulses at wavelength λ_s to be demultiplexed, optical sidebands satisfying the wavelength relationship: $1/\lambda_{FWM} = 2/\lambda_p - 1/\lambda_s$ are generated due to FWM in nonlinear materials. Note that the pump wavelength λ_p must coincide with the zero-dispersion wavelength of the fibers to satisfy the phase-matching condition of FWM process. If a repetition rate of the pump is one N-th of that for the original, one N-th, that is, 1-to-N demultiplexing can be fulfilled according to the pump pulses when the generated FWM pulses are filtered out from all output signals. Using 14-km dispersion-shifted fibers as a FWM device, the bit-error rates (BERs) of 16-Gbit/s to 4-Gbit/s demultiplexing were successfully measured [34]. This indicates the first error-free operation of all-optical DEMUX driven by LD sources with the aid of EDFAs.

Also the first error-free, polarization-independent (PI) 100-Gbit/s to 6.3-Gbit/s demultiplexing was demonstrated utilizing FWM in 3-km-long PANDA fibers with a polarization-rotating loop configuration, as shown in Figure 8.6 [35]. The configuration assures polarization independence, as orthogonal polarizations, namely S- and P-components, bidirectionally propagate the loop with their polarizations interchanged

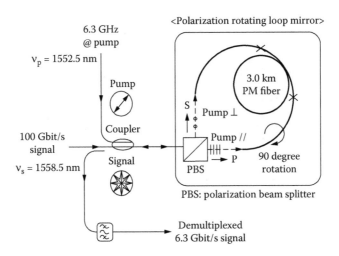

FIGURE 8.6 Polarization-independent DEMUX using FWM in polarization-rotating PANDA fiber loop mirror. (After Morioka [35].)

to each other. Since the polarization axis of the loop fiber is twisted by 90°, both components are lead to the input/output port through the polarization beam splitter. A polarization walk-off does not occur because both components use the same principal axis of the loop. To give the same FWM conversion efficiency for both components, the pump pulses are divided into S- and P-components with equal power. With this configuration, very stable polarization-independent (PI) demultiplexing was achieved with less than 1 pJ pumping and was used in the 100-Gbit/s, 200-km OTDM transmission experiment.

Furthermore, FWM-based multiple-output 100-Gbit/s demultiplexing was achieved employing a linearly chirped square pump pulse as shown in Figure 8.7 [36]. The instantaneous frequency difference between signal and pump pulses depends on their relative temporal position, and as a consequence, wavelength-converted pulses have different wavelengths according to the respective temporal position of the original 100-Gbit/s signal pulses. This experiment used the supercontinuum (SC)-pulse source followed by a chirp generation fiber as a chirped pump pulse generator. With this technique, 4-channel 10-Gbit/s signals were simultaneously demultiplexed from a 100-Gbit/s TDM signal.

As the fastest DEMNUX, 500-Gbit/s to 10-Gbit/s all-optical demultiplexing was demonstrated using FWM in a polarization-maintaining (PM) PANDA fiber [37]. Low-noise short SC pulses were used for both the 500-Gbit/s signal and the 10-GHz pump pulses. The SC for the pump was generated by pumping a 3-km single-mode dispersion-shifted fiber with 10-GHz 3.5-ps TL pulses from an Er-doped fiber ring-laser (EDFRL); the SC for the signal was generated by pumping a 1-km PANDA dispersion-shifted fiber with 10-GHz 3.5-ps TL pulses from another EDFRL. SCs were filtered by 3- to 5-nm optical bandpass filters (OBFs) to produce 1562.6-nm–0.98-ps signal pulses and 1548.9-nm–1.3-ps pump pulses. The root-mean-square (RMS) timing jitters of the pump and signal pulses were measured to be 77 fs and 97 fs, respectively.

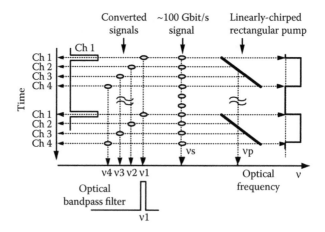

FIGURE 8.7 FWM-based multiple-output 100-Gbit/s demultiplexing employing a linearly chirped square pump pulse.

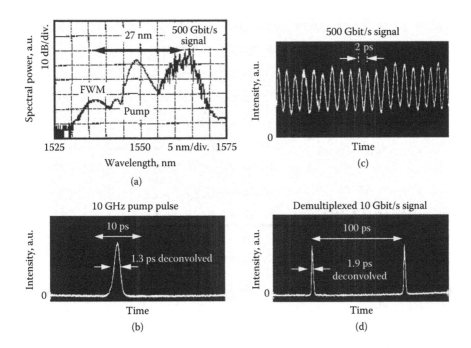

FIGURE 8.8 Spectrum and temporal waveform of 500-Gbit/s all-optical demultiplexing.

The signal pulses were then modulated at 10 Gbit/s and time-division-multiplexed by a factor of 50 in a PLC multiplexer to produce 500-Gbit/s TDM signals and were combined with the pump pulses in a WDM-MUX after amplification by EDFAs. The nonlinear medium for FWM was a 300-m PM-dispersion-shifted fiber (DSF) with a dispersion slope of 0.065 ps nm^{-2} km^{-1} and a zero-dispersion wavelength of 1549 nm. The combined pump and signal pulses were coupled into the PM-DSF with their polarization aligned along one of the principal axes. The generated FWM component was filtered out with two 5-nm OBFs in series. The demultiplexed signal was preamplified by an EDFA and its BER performance was measured as a function of the received power just before the EDFA. Figure 8.8 shows output spectra from the PM-DSF and temporal waveforms of the 500-Gbit/s signal stream, the 10-GHz pump pulse, and the demultiplexed 10-Gbit/s signal pulse. All the waveforms were measured with a newly developed optical sampling scope with a 0.5-ps temporal resolution described later (see Section 8.2.4). As shown here, the 10-Gbit/s signal pulses are clearly demultiplexed from the 500-Gbit/s signal, and the rms timing jitter was less than 0.1 ps.

FWM Switch in SOAs — Another prospective FWM-based DEMUX is using semiconductor optical amplifiers (SOAs) because of their compact size and lack of wavelength limitation as compared with FWM in fibers. The first error-free operation was reported using FWM in an SOA at 20-Gbit/s demultiplexing [38]. Since then lots of work has been reported using SOAs as a FWM device, such as 100-Gbit/s to 6.3-Gbit/s DEMUX [39], polarization-independent 200-Gbit/s to 6.3-Gbit/s

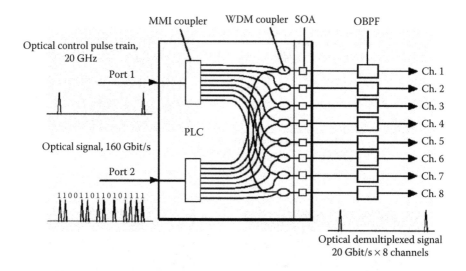

FIGURE 8.9 100-Gbit/s full optical time-division demultiplexing using FWM of SOA array integrated on PLC.

DEMUX [40], and other applications including phase-locked loop (PLL) based timing extraction described later. The polarization independence was fulfilled by using SOAs with small gain difference between the TE and TM modes, with the aid of depolarization of the input pump pulses using polarization walk-off of input PANDA fibers [40].

Recently, a noteworthy demonstration of full demultiplexing of 160 Gbit/s to 20 Gbit/s, applicable to real OTDM transmission systems, has been reported using a newly developed SOA-array hybrid-integrated demultiplexer based on FWM in SOAs [41]. Figure 8.9 shows the basic configuration of the DEMUX. It consists of two multimode-interference (MMI) couplers, eight WDM couplers, 16 different path-length waveguides, and two SOA-array bars. The array bars, having four SOAs each, are assembled on a PLC platform to make a compact DEMUX module (135 × 40 mm²). A 20-GHz control pulse train and a 160-Gbit/s OTDM signal are launched into port 1 and 2, respectively, and both are divided into eight streams by respective MMI couplers. Each control pulse is combined with an OTDM signal by a WDM coupler, and both are then input into an SOA to generate FWM light. Since each path length is chosen to demultiplex a different channel in the PLC, we have only to adjust the timing difference at the two input ports. With this OTDM-DEMUX together with the QPM-LN-based OTDM-MUX [29] previously described, 160-km OTDM transmission including full multiplexing eight-channel, 20-Gbit/s signals to a single 160-Gbit/s signal, and full demultiplexing from 160 Gbit/s to eight-channel, 20-Gbit/s signals, have successfully been demonstrated without no bit-error-rate floor for all the channels [30].

XPM Switch — An optical switch based on induced-frequency shifts is a promising approach to create a simple DEMUX because interferometry is not needed. Using the cross-phase modulation (XPM)-induced frequency shift, a polarization-independent 1:4

DEMUX that can provide four simultaneous outputs with one operation was demonstrated in the previously mentioned polarization-rotating mirror configuration [42]. Also, all-LD DEMUX operation using the same frequency-shift technique was carried out at 40 Gbit/s with the help of high power Er-doped fiber amplifiers (EDFAs) [43].

A novel multiple-output DEMUX has been proposed and demonstrated utilizing XPM [44]. Its operation is based on local chirp compensation of a down-chirped wide clock pulse through the XPM induced by a signal pulse stream. Figure 8.10 shows the principle. A linearly down-chirped clock pulse train and a randomly modulated 100-Gbit/s signal stream are coupled and launched into a nonlinear optical medium of positive nonlinear index ($n_2 > 0$). Signal pulses at times t_1 and t_2 induce a nonlinear optical frequency shift on the down-chirped clock pulse at frequencies v_1 and v_2 through XPM (Figure 8.10a). Since the XPM-induced frequency shift exhibits up-chirping in the center region (Figure 8.10b), the down-chirped clock pulse is locally compensated in the time domain at t_1 and t_2, resulting in an increase in the frequency components v_1 and v_2 according to the modulation signal pulses, as shown in Figure 8.10c. Consequently, simultaneous TDM to WDM signal conversion

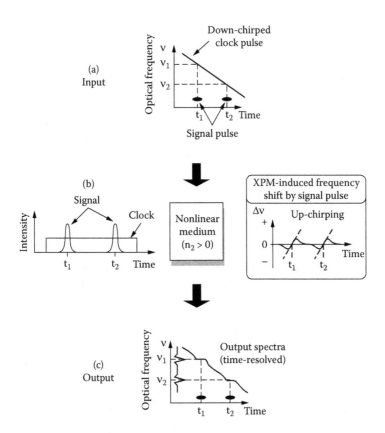

FIGURE 8.10 Principle of multi-output DEMUX using XPM-induced chirp compensation.

can be achieved by one operation. With this method, error-free, simultaneous 6-channel output, 100- to 6.3-Gbit/s demultiplexing was successfully demonstrated. Recently, full OTDM demultiplexing, from 100 Gbit/s to 10-channel, 10-Gbit/s signals, has successfully been demonstrated with this technique [45]. Since this method has a very simple setup compared to the conventional methods using interferometers or parametric processes, it may offer a practical solution to OTDM systems.

NOLM Switch — Another promising OTDM DEMUX configuration that uses induced-phase shift is a fiber-based Sagnac interferometer, usually called the nonlinear optical loop mirror (NOLM [46–54]). The operating principle of the NOLM is as follows: Incoming signal pulses are equally split into two counter-propagating pulses. They traverse the loop, recombine, and are normally back-reflected. If a control pulse propagates synchronously with one pulse, nonlinear effects introduce a phase shift to that particular pulse. At the NOLM exit, the phases of the corresponding pulse pair no longer match to assure back-reflection, but the combined pulse is partly directed to the NOLM output port. The back-reflected pulse is completely cancelled if the phase shift introduced amounts exactly to π.

As described above, since a NOLM utilizes the interference between two counter-propagating lights transiting the same fiber loop, there are, in principle, no residual phase differences between them. Accordingly, the NOLM switch can be of the order of several kilometers as compared with other interferometer-based switches, which leads to lower required pump powers; LD sources become feasible even if small nonlinearity materials such as silica fibers are used.

Following the primitive demonstration of a NOLM DEMUX [46,47], a 40-Gbit/s LD-pumped DEMUX using a 3-km dispersion-shifted (DS) fiber loop was reported [48], and the first error-free operation using a 14-km loop was confirmed at up to 64 Gbit/s [49]. After that, many NOLM experiments were reported including polarization-independent operation at 20 Gbit/s using dual-wavelength control pulses [51] and 100-Gbit/s to 6.3-Gbit/s polarization-independent DEMUX using a cross-spliced polarization-maintaining PANDA fiber loop [52]. Furthermore, multiple-channel output operation (1:5 DEMUX) was also dem-onstrated at 100 Gbit/s using a linearly chirped clock pulse as a signal port in the NOLM [53]. Lately, 640-Gbit/s DEMUX operation has been achieved by a specially designed NOLM DEMUX using dispersion-flattened fibers with no walk-off between signal and pump pulses [54]. This is the fastest operation speed at present and has been applied to the latest 640-Gbit/s single-channel OTDM transmission.

Figure 8.11 shows the polarization independent (PI) NOLM DEMUX [52]. It utilizes a PANDA fiber loop cross-spliced in the middle to cancel the overall polari-zation dispersion, resulting in no pulse walk-off when the signal pulse transits the 3-km-long loop. Since the control pulse equally excites two principal axes, two sets of single-polarization NOLMs can be formed corresponding to the orthogonal polar-ization states. This method yielded the first error-free polarization-independent oper-ation at 32 to 6.3 Gbit/s and 100 to 6.3 Gbit/s demultiplexing.

Figure 8.12 illustrates multiple-output DEMUX using a NOLM switch [53]. It can be realized by using the largely chirped supercontinuum (SC) pulse as a signal

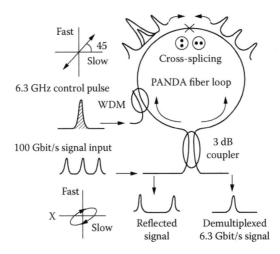

FIGURE 8.11 Polarization-independent NOLM switch.

to be demultiplexed and the input OTDM signal as a pump. The different colors are switched from the chirped SC pulse according to the relative temporal position between the SC pulse and the input OTDM signal, thus leading to the OTDM to WDM conversion. With this configuration, 5-channel 10-Gbit/s signals were demultiplexed from a 100-Gbit/s TDM signal at the same time.

TOAD and SLALOM — The first demultiplexing experiment using cross-phase modulation in SOAs was performed in a hybrid arrangement of a fiber loop mirror (NOLM) with a SOA called SLALOM [55] or TOAD [57]. The intention of these first experiments was a demonstration of the switching principle. Later, these hybrid switching devices were used in more advanced system experiments in order

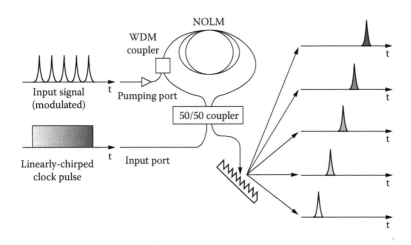

FIGURE 8.12 Multi-output NOLM demultiplexer using linearly chirped SC pulses.

to demultiplex to a low bit-rate channel (e.g., 10 Gbit/s) from an optically-multiplexed high-bit-rate data signal of up to 160 Gbit/s [56,58,59].

It is necessary for the SOA-based switching to circumvent the speed limitation due to relatively slow carrier recovery time in SOAs, leading to slow turn-off time. With offsetting a SOA from the center of the loop, the TOAD arrangement can utilize only the phase difference between two counter-propagating probe lights transiting the same SOA with a delay time. Thus the switching window of the TOAD is mainly determined by the delay time, independent of the slow recovery time of overall refractive index change in the SOA.

Recently, making full use of the above features, SOA-based hybrid-integrated interferometric devices have been realized utilizing differential phase modulation, DPM in short, between two SOAs (Mach-Zehnder interferometer [60,61]) or between two counter-propagating lights through one SOA (Sagnac interferometer [62]). The operation speed of error-free demultiplexing has been improved to 168 Gbit/s [60] and 336 Gbit/s [61] using a symmetric Mach-Zehnder type switch. Also, filter-free wavelength conversion or bit-rate and format conversion operation has been demonstrated at 40 Gbit/s using a TOAD structure device, called SIPAS, in which all the elements, such as SOA, loop waveguides and 3-dB couplers, are monolithically integrated on a semiconductor platform [62].

Future Trends — Recent studies on demultiplexing technique have been focused on practical applications aiming at next 160- to 320-Gbit/s transmission systems using SOAs, highly nonlinear materials, or various O/O-IC or O/E-IC technology. For instance, a 160-Gbit/s to 10-Gbt/s OTDM demultiplexer based on XPM-induced wavelength shifting has been demonstrated using highly nonlinear fiber [63]. Also, fast electro-optical demultiplexing using a monolithic PD-EAM optical gate has been successfully demonstrated at 320 Gbit/s and 500 Gbit/s [64]. This utilizes an EA modulator gate driven by output voltage of fast UTC-PD which is triggered by ultrashort optical pulses. Furthermore, 320-Gbit/s demultiplexing using a quasi–phase-matched $LiNbO_3$ (QPM-LN) device has been reported; this device works by offsetting the group delay difference between the fundamental and second harmonic (SH) lights in advance [65].

8.2.3 OPTICAL TIMING EXTRACTION

Timing extraction, which extracts the timed clock from the received optical signals, is one of the key functions for constructing high-speed optical transmission systems. All transmission apparatuses such as repeaters and demultiplexers require accurate clock extraction. Requirements for optical timing extraction include ultrafast operation, low phase noise, high sensitivity, and polarization independence. In particular, ultralow phase noise yielding <1-ps rms jitter is required to achieve 160-Gbit/s transmission. The highest operating speed of currently used electrical timing extraction circuits is as high as 40 GHz, determined by the microwave mixer used as a phase detector in a phase-lock loop (PLL) circuit [66]. To increase the operating speed, many approaches based on photonic technology have been studied, as shown in Table 8.3. They are classified into three categories: optical tank circuits using a Fabry-Perot resonator [67] or Brillouin gain in fiber [68], injection locking of self-pulsating

TABLE 8.3
High-Speed Timing Extraction Techniques

Principle		Speed (Gbit/s)	Author	Year	Ref.
Optical tank circuit	Fabry-Perot etalon	2	Jinno (NTT)	1992	67
	Brillouin gain in fiber	5	Miyamoto (NTT)	1993	68
Injection locking	Self-pulsating DFB-LD	0.2	Jinno (NTT)	1988	69
		5	Barnsley (BT)	1991	70
		18	As (HHI)	1993	71
		10	Sartorius (HHI)	1997/98	72
	Phase COM Laser	40/80	Bornholdt (HHI)	99/2002	73/74
	Mode-locked LD	10	Takayama (NTT)	1990	75
		5	Ono (NTT)	1994	76
		10	Ono (NEC)	1995	77
		80	Miyazaki (CRL)	2002	78
		100	Ohno (NTT)	2002	79
	Mode-locked fibre laser	1	Smith (BT)	1992	80
		40	Ellis (BT)	1993	81
		20	Patrick (BT)	1994	82
Phase Lock Loop (PLL)	Electrical PLL	40	Ellis (BT)	1993	66
	$LiNbO_3$ modulator	14	Takayama (NTT)	1992	83
	Gain mod. in TW-SOA	1	Kawanishi (NTT)	1988	84
		50	Kawanishi (NTT)	1993	85
	FWM in TW-SOA	50	Kamatani (NTT)	1994	86
		100	Kawanishi (NTT)	1994	87
		200/400	Kamatani (NTT)	1994/95	88
	XPM (SOA in SLALOM)	160	Yamamoto (HHI)	2002	90
	EA modulator	160	Boerner (HHI)	2003	91
	Cascade SHG (QPM-LN)	160	Ohara (NTT)	2003	92
	FWM in fiber	20	Saito (NEC)	1993	89

Note: TW-SOA = traveling-wave semiconductor-optical amplifier; FWM = four-wave mixing.

LDs [69–74], mode-locked LDs [75–79], EDF lasers [80–82], and phase-lock loop circuits using electrical voltage-controlled oscillators (VCOs) [83–92].

An optical tank circuit features a simple configuration originating from its passive structure. The optical signal created by return-to-zero intensity modulation consists of a continuous component and line components in an optical spectrum. Timing extraction can be achieved by optically extracting the line spectrum components spacing at clock frequency. Although an optical clock was extracted from 2-Gbit/s pseudo-random optical data using a Fabry-Perot resonator [67], it is difficult to stably operate due to its sensitivity to temperature and vibration.

Injection locking utilizes self-pulsating (SP) LDs, such as multielectrode LDs [69–72] or optical inverters which alternate between TE and TM modes [75], whose output repetition frequency is locked to that of an injected optical pulse train. Using these techniques, 5- and 10-GHz clocks were generated all-optically. A wavelength- and polarization-insensitive clock recovery module based on an SP DFB-LD was also developed and tested in a 10-Gbit/s, 105-km transmission experiment [72]. Recently, very high-speed all-optical clock recovery of 40 GHz from 40-Gbit/s signals or 80 GHz from 80-Gbit/s signals has been demonstrated using newly developed self-pulsating DFB-LDs called Phase COM LDs [73,74]. Phase COM LDs can simultaneously output two line-spectrum components, thus leading to the sinusoidal intensity modulation at beat frequency. With this technique, an 80-Gbit/s, 160-km transmission experiment was successfully conducted [74].

Clock recovery using injection locking of mode-locked LDs or mode-locked fiber lasers was limited to 10 GHz or 40 GHz at most [77,81]. Recently it has been found that the subharmonic injection locking may occur using mode-locked LDs. In other words, 10-GHz mode-locked LDs can be locked by injecting N-times-10 (where N is an integer) Gbit/s OTDM signals. With this technique, optical sampling and all-optical demultiplexing of 80-Gbit/s OTDM signals have successfully been demonstrated by recovering a subharmonic 10-GHz clock from 80-Gbit/s signals [78]. Also, 40-GHz optical clock recovery from a 160-Gbit/s optical data stream was successfully demonstrated using a regeneratively mode-locked semiconductor laser [79]. A subharmonic injection technique has a prospective feature capable of directly extracting the divided clock frequency all-optically from very high-speed OTDM signals.

An all-optical signal processor, which used an all-optical clock recovery scheme based on a mode-locked Er fiber laser [80–82] and a NOLM regenerator [93], has also been demonstrated. It restored the pulse timing and removed the noise and intensity fluctuations [94]; however, its performance so far has been insufficient. Nevertheless, this technique will be applied to very long 1R repeated transmission systems if its performance can be improved.

A PLL has several advantages in that it has no phase error and complete retiming is, in principle, possible. The question is how fast the PLLs can operate. Because the operational speed of conventional electric PLL circuits is limited by the response of the phase comparators used, new cross-correlation techniques based on photonics have been developed to overcome this problem. Figure 8.13 shows the principle of a PLL circuit in which an SOA is used as an all-optical phase correlator [84]. Because the SOA gain is instantaneously modulated by the intense optical clock, the cross-correlation (Δf component) between the 10-Gbit/s signal and the (10 GHz + Δf) clock driven by a voltage-controlled oscillator (VCO) can be obtained all-optically. The cross-correlated Δf signal, which contains the phase difference between the optical signal and optical clock, is compared with the reference Δf signal in a conventional low-speed electrical phase comparator, and the output is returned to the VCO to close the PLL. With this PLL configuration, a 6.3-GHz retimed signal has been recovered using a residual 6.3-GHz component in a 100-Gbit/s optical TDM signal.

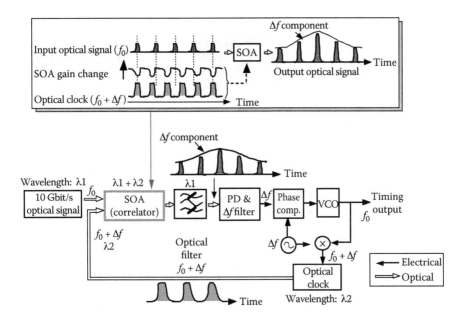

FIGURE 8.13 Principle of PLL based on an all-optical SOA correlator.

In the case of completely multiplexed TDM signals, there is no low-frequency clock component capable of being extracted. To extract the retimed clock, the phase detection must be processed at the multiplexed clock frequency, that is, 100 GHz for a 100-Gbit/s signal. By applying a PLC multiplier, which generates an N-times multiplied optical clock frequency, to the above PLL, the original 6.3-GHz clock has been extracted from a completely multiplexed 50-Gbit/s signal [85]. This indicates that the SOA phase detector has adequate response to operate at 50 GHz or beyond.

A prescaled PLL circuit, which extracts the prescaled frequency clock of 6.3 GHz (1/16th of 100 GHz) from a randomly modulated 100-Gbit/s signal, has also been demonstrated [86,87]. Figure 8.14 shows the configuration of the prescaled PLL. This utilizes the harmonic frequency components that the short optical clock possesses and the high-speed four-wave-mixing process in the SOA instead of gain saturation. A prescaled 6.3-GHz clock was recovered from a 100-Gbit/s TDM optical signal using the 100-GHz phase comparison output obtained as the cross-correlation ($16 \times \Delta f$ component) between a 100-Gbit/s optical signal and a short (< 10 ps) 6.3 GHz + Δf optical clock pulse train in the FWM light; clock pulse multiplication is not required. The rms jitter is measured to be less than 0.3 ps. In order to increase the correlation signal level, pulses of less than 1-ps duration from the SC sources are used for the short optical clock pulse source and the OTDM signal source as well. With this method, the prescaled frequency clock of 6.3 GHz (1/64th of 400 GHz) can be extracted from a randomly modulated 400-Gbit/s signal [88]. This method has been applied to the latest 400-Gbit/s TDM and 1- to 3-Tbit/s TDM/WDM transmission experiments.

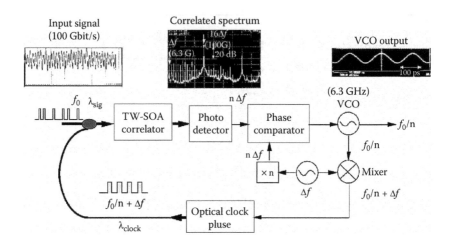

FIGURE 8.14 Configuration of prescaled PLL.

Recently, various timing extraction techniques have extensively been studied, aiming at the next 160 Gbit/s optical transmission systems. They are all based on the PLL circuit, but the difference is only the correlation detection technique used. A 40-GHz optical clock extraction from 160-Gbit/s data signals has been successfully demonstrated using a SLALOM or TOAD configuration as balanced phase detection of PLL [90]. A 10-GHz clock recovery from a 160-Gbit/s data signal with an electro-optic PLL using a bidirectionally operated electroabsorption (EA) modulator as a phase comparator has also been reported [91]. Furthermore, a 20-GHz clock has been recovered from a 160-Gbit/s data signal by using a PLL with an optical phase modulator for sideband generation of a local optical clock, and with periodically poled lithium niobate (PPLN) for cascade SHG for correlated optical signal generation [92].

8.2.4 HIGH-SPEED OPTICAL WAVEFORM MEASUREMENT

Optical waveform measurement with high temporal resolution is one of the key techniques in developing ultrahigh-speed OTDM transmission systems. In particular, eye diagram measurements, which measure variations or fluctuations in randomly modulated optical waveforms, are needed to evaluate very high-speed optical transmission characteristics. Optical sampling, where optical waveforms to be measured are all-optically sampled (gated) by intense short optical pulses, provides a powerful tool for measuring high-speed optical waveforms [95–103]. Several all-optical gates have been studied to perform the necessary cross-correlation between sampling and signal pulses. These include sum-frequency generation (SFG) in nonlinear crystals, XPM in fibers or SOAs, and FWM in fibers or SOAs. Here, recent SFG-based optical sampling having less than 1-ps resolution will be described together with its effective data.

Figure 8.15 compares the principle of optical sampling and conventional electrical sampling used for optical waveform measurements. Temporal resolution of the conventional approach is limited by the photodetector bandwidth as well as that of the electrical sampler used. Optical sampling can circumvent the electronics limitation and its

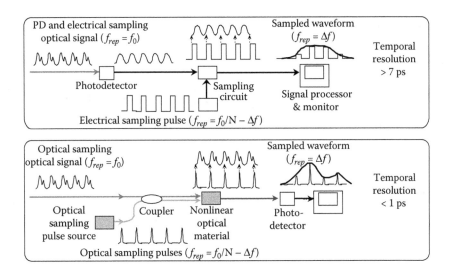

FIGURE 8.15 Principle of optical sampling.

resolution is determined by the sampling pulse width, provided that the overall nonlinear response is sufficiently fast. This is the case for short SFG crystals. Also, the timing jitter of the sampling pulse should be minimized to achieve best temporal resolution of less than 0.6 ps [95]. Note that the timing jitter of the signal pulse stream should be evaluated itself; thus the conventional autocorrelation technique using second-harmonic generation (SHG) cannot be adopted in the case under consideration here.

Figure 8.16 depicts the experimental setup of optical sampling used for eye-diagram measurements of a 100-Gbit/s optical signal [96]. For the signal source, 10-GHz, 4.4-ps optical pulses from a stabilized mode-locked EDF laser were used. After modulation by a $LiNbO_3$ optical modulator, the 10-Gbit/s optical signal was multiplexed to a 100-Gbit/s OTDM signal with a PLC-MUX inserted between two EDFAs. The 100-Gbit/s optical signal and the 10-GHz electrical clock f_0 were input into the optical sampling system. The timing clock generator generated electrical timing clocks of $f_0 - \Delta f$ and $(f_0 - \Delta f)/1024$. These clocks were used for generating nearly 10-GHz optical pulses from the second EDF laser and for extracting every 1024th pulse from the 10-GHz pulses by the second modulator, respectively. The extracted 9.7-MHz optical pulses were compressed to 0.4-ps duration by the SC technique and amplified to a 230-W peak by the EDFA. Both the 100-Gbit/s signal and the optical sampling pulses were combined with a polarization beam splitter and coupled into the organic AANP (2-adamantylamino-5-nitropyridine) crystal for type-II phase-matching SFG. The cross-correlation signals (SF light) at $(f_0 - \Delta f)/1024$ (= 9.7 MHz) were detected with an Si-APD and the obtained electrical signals were processed by a computer to be displayed on a monitor.

Figure 8.17 compares the waveforms of the 160-Gbit/s optical signal as measured by a streak camera and the proposed optical sampling system, respectively. It is apparent that optical sampling can accurately measure the eye diagram of the 160-Gbit/s optical signal, whereas the streak camera shows the all-mark rough pattern such as

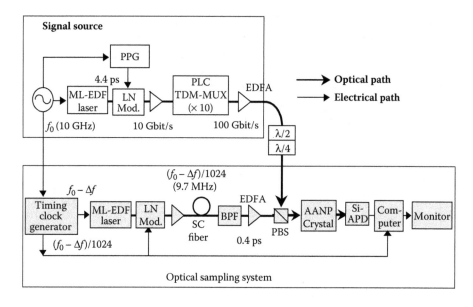

FIGURE 8.16 Experimental setup of optical sampling system.

{1,1,1,1} because only the averaged waveforms could be measured at 1.5-μm due to its low sensitivity and limited response.

Recently, a lot of attention has been paid to ultrafast optical sampling, and very fast eye-diagram measurement has been reported using various methods. They include the first 300-Gbit/s eye-diagram measurement by optical sampling using parametric amplification (FWM) in short (50 m) highly nonlinear fibers (HNLFs) and a 1.6-ps optical sampling pulse [97]; 160-Gbit/s eye-diagram measurement with the NOLM gate, using a HNLF together with a SLALOM gate-based clock recovery PLL [98]; and 320-Gbit/s eye-diagram measurement using a KTP SHG crystal together with a very stable, passively mode-locked fiber laser pulse with 700-fs time resolution [99].

Lately, much faster optical sampling experiments have been demonstrated, including 500-Gbit/s eye-diagram measurements utilizing XPM-induced wavelength

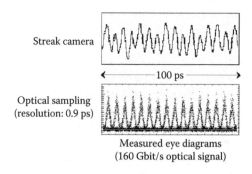

FIGURE 8.17 160-Gbit/s eye diagram measured with optical sampling.

shifting in a 50-m HNLF [100] and 640-Gbit/s eye-diagram measurements using a PPLN SFG crystal together with a wavelength-tunable 210-fs soliton pulse [101]. In addition, optical sampling systems have also been applied for on-line optical signal monitoring by actively measuring a Q factor of 160 Gbit/s OTDM signal after transmission with synchronous sampling [102] or asynchronous sampling [103].

8.3 APPLICATIONS TO VERY HIGH-SPEED OTDM AND OTDM/WDM TRANSMISSION [104]

In 1993, utilizing newly developed key technologies including TL pulse sources, MUX/DEMUX circuits, and timing extraction PLLs, the first successful 100-Gbit/s-50-km transmission was conducted by eightfold-OTDM together with twofold polarization multiplexing (2-PDM). Since then, OTDM transmission technologies have been made a great deal of progress toward much faster and longer optical transmission systems. By improving and developing key technologies, especially a wavelength-selectable subpicosecond pulse source based on supercontinuum (SC) generation and a FWM-based all-optical DEMUX together with a prescaled PLL, 200- Gbit/s–100-km transmission, and 400-Gbit/s–40-km transmission were achieved using the OTDM technique. Also, a single-wavelength channel 640-Gbit/s OTDM transmission experiment over 60 km was conducted using a 640-fs pulse train from a mode-locked EDF laser and walk-off free, dispersion-flattened NOLM demultiplexer. Furthermore, by utilizing twofold polarization-multiplexing together with higher-order dispersion compensation, 1.28-Tbit/s single-channel OTDM transmission over 70 km was successfully conducted with the same setup.

The construction of large-capacity flexible optical networks, using both OTDM and WDM technologies, will be of vital importance. In these networks, broadband low-noise optical sources, such as SC pulse sources, are expected to play a major role. In 1995, the first OTDM/WDM transmission experiment with a 400-Gbit/s total capacity was demonstrated using 100 Gbit/s × 4 WDM channels generated from a single SC pulse source. This experiment utilized the striking feature of the SC source: it could generate short pulses less than 0.3 ps over the continuous spectral range (200 nm), and multi-wavelength transform-limited short pulses could easily be selected by filtering with passive optical filters. Because the optical frequency characteristics, including frequency separation and stability, are determined by the filtering devices, it is easy to generate dense WDM pulses from a single SC source. Using an AWG filter as a WDM demultiplexer/multiplexer, the first 1-Tbit/s capacity (100 Gbit/s × 10 WDM channels) OTDM/WDM transmission over 40 km was successfully demonstrated in 1996.

More recently, 3-Tbit/s (160 Gbit/s × 19 channels) OTDM/WDM transmission through 40 km of dispersion-shifted fiber was reported, employing two sets of low-noise, flatly broadened SC WDM pulse sources and ultra-broadband EDFAs. The 3-Tbit/s signal generator consisted of two OTDM/WDM signal generators one for the shorter wavelength region and one for the longer wavelength region. With this setup, 3-Tbit/s error-free transmission through a 40-km DS fiber was successfully demonstrated utilizing both 16-fold OTDM and 19-fold WDM techniques.

REFERENCES

1. A. Takada, T. Sugie, and M. Saruwatari, "High-speed picosecond optical pulse compression from gain-switched 1.3-μm distributed feedback-laser diode (DFB-LD) through highly dispersive single-mode fiber," *IEEE J. Lightwave Technol.*, LT-5, 1525–1533, 1987.

2. M. Suzuki, H. Taga, H. Tanaka, N. Edagawa, K. Utaka, S. Yamamoto, Y. Matsushima, K. Sakai, and H. Wakabayashi, "Transform-limited optical pulse generation up to 20 GHz repetition rate by sinusoidally driven InGaAsP electroabsorption modulator," *Conf. Lasers and Electro-Optics*, (CLEO '92), postdeadline paper CPD 26, 1992.

3. K. Suzuki, K. Iwatsuki, S. Nishi, M. Saruwatari, K. Sato, and K. Wakita, "2.5 ps soliton pulse generation at 15 GHz with monolithically integrated MQW-DFB-LD/MQW-EA modulator and dispersion decreasing fiber," *Techn. Digest of Optical Amplifiers and their Applications*, OSA Techn. Digest Series 14, Opt. Soc. America, Washington, DC, 1993, p. 314.

4. Y.K. Chen, M.C. Wu, T. Tanbun-Ek, R.A. Logan, and M.A. Chin, "Subpicosecond monolithic colliding-pulse mode-locked multiple quantum-well lasers," *Appl. Phys. Lett.*, 58, 1253–1255, 1991.

5. P.B. Hansen, G. Raybon, U. Koren, B.I. Miller, M.G. Young, M. Chien, C.A. Burrus, and R.C. Alferness, "5.5-mm long InGaAsP monolithic extended-cavity laser with an integrated Bragg-reflector for active mode-locking," *IEEE Photon. Techol. Lett.*, 4, 215–217, 1992.

6. A. Takada, K. Sato, M. Saruwatari, and M. Yamamoto, "Pulse width tunable subpicosecond pulse generation from an actively modelocked monolithic MQW laser/MQW electroabsorption modulator," *Electron. Lett.*, 30, 898–900, 1994.

7. Y. Katagiri, A. Takada, S. Nishi, H. Abe, Y. Uenishi, and S. Nagaoka, "Passively mode-locked micromechanically-tunable semiconductor lasers," *IEICE Trans. Electron.*, E81-C, 151–159, 1998.

8. R. Ludwig, S. Diez, A. Ehrhardt, L. Küller, W. Pieper, and H.G. Weber, "A tunable femtosecond modelocked semiconductor laser for applications in OTDM-systems," *IEICE Trans. Electron.*, E81-C, 140–145, 1998.

9. S. Arahira, S. Kutsuzawa, Y. Matsui, and Y. Ogawa, "Higher order chirp compensation of femtosecond mode-locked semiconductor lasers using optical fibers with different group-velocity dispersions," *IEEE J. Select. Topics Quantum Electron.*, 2, 480–486, 1996.

10. K. Sato, I. Kotaka, A. Hirano, M. Asobe, Y. Miyamoto, N. Shimizu, and K. Hagimoto, "High-repetition frequency pulse generation at 102 GHz using mode-locked lasers integrated with electroabsorption modulators," *Electron. Lett.*, 34, 790–792, 1998.

11. A. Takada and H. Miyazawa, "30 GHz picosecond pulse generation from actively mode-locked erbium-doped fibre laser," *Electron. Lett.*, 26, 216–217, 1990.

12. G.T. Harvey and L.F. Mollenauer, "Harmonically mode-locked fiber ring laser with an internal Fabry-Perot stabilizer for soliton transmission," *Opt. Lett.*, 18, 107–109, 1993.

13. X. Shan and D.M. Spirit, "Novel method to suppress noise in harmonically mode-locked erbium fibre lasers," *Electron. Lett.*, 29, 979–981, 1993.

14. H. Takara, S. Kawanishi, M. Saruwatari, and K. Noguchi, "Generation of highly stable 20 GHz transform-limited optical pulses from actively mode-locked Er^{3+}-doped fibre lasers with an all-polarisation maintaining ring cavity," *Electron. Lett.*, 28, 2095–2096, 1992.

15. H. Takara, S. Kawanishi, and M. Saruwatari, "20 GHz transform-limited optical pulse generation and bit-error-free operation using a tunable, actively modelocked Er-doped fibre ring laser," *Electron. Lett.*, 29, 1149–1150, 1993.

16. H. Takara, S. Kawanishi, and M. Saruwatari, "Stabilisation of a modelocked Er-doped fibre laser by suppressing the relaxation oscillation frequency component," *Electron. Lett.*, 31, 292–293, 1995.

17. E. Yoshida and M. Nakazawa, "Wavelength tunable 1.0 ps pulse generation in 1.530-1.555 μm region from PLL, regeneratively modelocked fibre laser," *Electron. Lett.*, 34, 1753–1754, 1998.

18. K. Mori, T. Morioka, H. Takara, and M. Saruwatari, "Continuously tunable optical pulse generation utilizing supercontinuum in an optical fiber pumped by an amplified gain-switched LD pulses," *Techn. Digest of Optical Amplifiers and their Applications*, OSA Techn. Digest Series 14, Opt. Soc. America, Washington, DC, 1993, p. 190.

19. T. Morioka, S. Kawanishi, K. Mori, and M. Saruwatari, "Nearly penalty-free, <4 ps supercontinuum Gbit/s pulse generation over 1535-1560 nm," *Electron. Lett.*, 30, 790–791, 1994.

20. K. Mori, H. Takara, S. Kawanishi, M. Saruwatari, and T. Morioka, "Flatly broadened supercontinuum spectrum generated in a dispersion decreasing fibre with convex dispersion profile," *Electron. Lett.*, 33, 1806–1808, 1997.

21. H. Takara, T. Ohara, K. Mori, K. Sato, E. Yamada, K. Jinguji, Y. Inoue, T. Shibata, T. Morioka, and K.-I. Sato, "Over 1000 channel optical frequency chain generation from a single supercontinuum source with 12.5 GHz channel spacing for DWDM and frequency standards," ECOC 2000, PDP 3.1.

22. K. Mori, K. Sato, H. Takara, and T. Ohara, "Supercontinuum lightwave source generating 50 GHz spaced optical ITU grid seamlessly over S-, C- and L-bands," *Electron. Lett.*, 39, 544–545, 2003.

23. S. Kawanishi and O. Kamatani, "All-optical time division multiplexing using four-wave mixing," *Electron. Lett.*, 30, 1697–1698, 1994.

24. S. Kawanishi, K. Okamoto, M. Ishii, O. Kamatani, H. Takara, and K. Uchiyama, "All-optical time-division-multiplexing of 100 Gbit/s signal based on four-wave mixing in a travelling-wave semiconductor laser amplifier," *Electron. Lett.*, 33, 976–977, 1997.

25. I. Shake, H. Takara, S. Kawanishi, and M. Saruwatari, "High-repetition-rate optical pulse generation by using chirped optical pulses," *Electron. Lett.*, 34, 792–793, 1998.

26. H. Takara, I. Shake, K. Uchiyama, O. Kamatani, S. Kawanishi, and K. Sato, "Ultra-high-speed optical TDM signal generator utilizing all-optical modulation and optical clock multiplication," *Conf. Opt. Fiber Commun. (OFC '98)*, OSA Techn. Digest Series, Opt. Soc. America, Washington, DC, 1998, postdeadline paper PD16.

27. S. Kawanishi, Y. Miyamoto, H. Takara, M. Yoneyama, K. Uchiyama, I. Shake, and Y. Yamabayashi, "120 Gbit/s OTDM system prototype," Postdeadline Papers of 24th European Conf. on Optical Communication (ECOC 1998), p. 41.

28. H. Murai, M. Kagawa, H. Tsuji, K. Fujii, Y. Hashimoto, and H. Yokoyama, "Single channel 160 Gbit/s (40 Gbit/s × 4) 300 km – transmission using EA modulator based-OTDM module and 40 GHz external-cavity mode-locked LD," ECOC '02, No. 2.1.4.

29. T. Yamada, M. Ishii, S. Mino, T. Shibata, T. Ohara, H. Takara, I. Shake, and S. Kawanishi, "Compact and highly efficient optical-time-division-multiplexer using low loss direct attachment with 1.5% DPLC and PPLN waveguides," IEEE LEOS Annual Meeting, WFF1, 2003.

30. T. Ohara, H. Takara, I. Shake, K. Mori, K. Sato, S. Kawanishi, S. Mino, T. Yamada, M. Ishii, I. Ogawa, T. Kitoh, K. Magari, M. Okamoto, R.V. Roussev, J.R. Kurz, K.R. Parameswaran, and M.M. Fejer, "160-Gb/s OTDM transmission using integrated all-optical MUX/DEMUX with all-channel modulation and demultiplexing," *IEEE Photon. Technol. Lett.*, 16, 650–652, 2004.

31. T. Morioka, M. Saruwatari, and A. Takada, "Ultrafast optical multi/demultiplexer utilising optical Kerr effect in polarisation-maintaining single-mode fibres," *Electron. Lett.*, 23, 453–454, 1987.

32. T. Morioka, H. Takara, K. Mori, and M. Saruwatari, "Ultrafast reflective optical Kerr demultiplexer using polarisation rotation mirror," *Electron. Lett.*, 28, 521–522, 1992.

33. D.M. Patrick and A.D. Ellis, "Demultiplexing using crossphase modulation-induced spectral shifts and Kerr polarisation rotation in optical fibre," *Electron. Lett.*, 29, 227–229, 1993.

34. P.A. Andrekson, N.A. Olsson, J.R. Simpson, T. Tanbun-Ek, R.A. Logan, and M. Haner, "16 Gbit/s all-optical demultiplexing using four-wave mixing," *Electron. Lett.*, 27, 922–924, 1991.

35. T. Morioka, S. Kawanishi, K. Uchiyama, H. Takara, and M. Saruwatari, "Polarisation-independent 100 Gbit/s all-optical demultiplexer using four-wave mixing in a polarisation-maintaining fibre loop," *Electron. Lett.*, 30, 591–592, 1994.

36. T. Morioka, S. Kawanishi, H. Takara, and M. Saruwatari, "Multiple output, 100 Gbit/s all-optical demultiplexer based on multichannel four-wave mixing pumped by a linearly-chirped square pulse," *Electron. Lett.*, 30, 1959–1960, 1994.

37. T. Morioka, H. Takara, S. Kawanishi, T. Kitoh, and M. Saruwatari, "Error-free 500 Gbit/s all-optical demultiplexing using low-noise, low-jitter supercontinuum short pulses," *Electron. Lett.*, 32, 833–834, 1996.

38. R. Ludwig and G. Raybon, "All-optical demultiplexing using ultrafast four-wave-mixing in a semiconductor laser amplifier at 20 Gbit/s," *Proc. 19th Europ. Conf. Opt. Commun. (ECOC '93)*, 3, 57 (1993).

39. S. Kawanishi, T. Morioka, O. Kamatani, H. Takara, J.M. Jacob, and M. Saruwatari, "100 Gbit/s all-optical demultiplexing using four-wave mixing in a travelling wave laser diode amplifier," *Electron. Lett.*, 30, 981–982, 1994.

40. T. Morioka, H. Takara, S. Kawanishi, K. Uchiyama, and M. Saruwatari, "Polarisation-independent all-optical demultiplexing up to 200 Gbit/s using four-wave mixing in a semiconductor laser amplifier," *Electron. Lett.*, 32, 840–842, 1996.

41. I. Shake, H. Takara, K. Uchiyama, I. Ogawa, T. Kitoh, T. Kitagawa, M. Okamoto, K. Magari, Y. Suzuki, and T. Morioka, "160 Gbit/s full optical time-division demultiplexing using FWM of SOA-array integrated on PLC," *Electron. Lett.*, 38, 37–38, 2002.

42. T. Morioka, K. Mori, and M. Saruwatari, "Ultrafast polarisation-independent optical demultiplexer using optical carrier frequency shift through crossphase modulation," *Electron. Lett.*, 28, 1070–1072, 1992.

43. D.M. Patrick and A.D. Ellis, "Demultiplexing using crossphase modulation-induced spectral shifts and Kerr polarisation rotation in optical fibre," *Electron. Lett.*, 29, 227–229, 1993.

44. K. Uchiyama, S. Kawanishi, and M. Saruwatari, "Multiple-channel output all-optical demultiplexer based on TDM-WDM conversion utilizing time-division chirping control of chirped clock pulse," *Proc. 23rd Europ. Conf. Opt. Commun. (ECOC '97)*, 3, 71 (1997);

45. K. Uchiyama, "All-optical signal processing for 160 Gbit/s/channel OTDM/WDM systems," OFC 2001, Anaheim, CA, March 22, 2001, ThH2.

46. K.J. Blow, N.J. Doran, and B.P. Nelson, "Demonstration of the nonlinear fibre loop mirror as an ultrafast all-optical demultiplexer," *Electron. Lett.*, 26, 962–964, 1990.

47. M. Jinno and T. Matsumoto, "Ultrafast low-power and highly stable fiber Sagnac interferometer," *IEEE Photon. Technol. Lett.*, 2, 349–351, 1990.

48. A. Takada, K. Aida, and M. Jinno, "Demultiplexing of a 40 Gb/s optical signal to 2.5 Gb/s using a nonlinear fiber loop mirror driven by amplified, gain-switched laser diode pulses," *Conf. Opt. Fiber Commun. (OFC '91)*, OSA Techn. Digest Series, Opt. Soc. America, Washington, DC, 1991, p. 50.

49. P.A. Andrekson, N.A. Olsson, J.R. Simpson, D.J. Digiovanni, P.A. Morton, T. Tanbun-Ek, R.A. Logan, and K.W. Wecht, "Ultrahigh speed demultiplexing with the nonlinear optical loop mirror," *Conf. Opt. Fiber Commun. (OFC '92)*, OSA Techn. Digest Series, Opt. Soc. America, Washington, DC, 1992, postdeadline papers, 343.

50. D.M. Patrick, A.D. Ellis, and D.M. Spirit, "Bit-rate flexible all-optical demultiplexing using a nonlinear optical loop mirror," *Electron. Lett.*, 29, 702–703, 1993.

51. H. Bülow and G. Veith, "Polarisation-independent switching in a nonlinear optical loop mirror by a dual-wavelength switching pulse," *Electron. Lett.*, 29, 588–589, 1993.

52. K. Uchiyama, S. Kawanishi, H. Takara, T. Morioka, and M. Saruwatari, "100 Gbit/s to 6.3 Gbit/s demultiplexing experiment using polarisation-independent nonlinear optical loop mirror," *Electron. Lett.*, 30, 873–874, 1994.

53. K. Uchiyama, H. Takara, T. Morioka, S. Kawanishi, and M. Saruwatari, "100 Gbit/s multiple-channel output all-optical demultiplexer based on TDM-WDM conversion in a nonlinear optical loop mirror," *Electron. Lett.*, 32, 1989–1991, 1996.

54. T. Yamamoto, E. Yoshida, and M. Nakazawa, "Ultrafast nonlinear optical loop mirror for demultiplexing 640 Gbit/s TDM signals," *Electron. Lett.*, 34, 1013–1014, 1998.

55. M. Eiselt, W. Pieper, and H.G. Weber, "All-optical high speed demultiplexing with a semiconductor laser amplifier in a loop mirror configuration," *Electron. Lett.*, 29, 1167–1168, 1993.

56. S. Diez, R. Ludwig, and H.G. Weber, "All-optical switch for TDM and WDM/TDM systems demonstrated in a 640 Gbit/s demultiplexing experiment," *Electron. Lett.*, 34, 803–805, 1998.

57. J.P. Sokoloff, P.R. Prucnal, I. Glesk, and M. Kane, "A terahertz optical asymmetric demultiplexer (TOAD)," *IEEE Photon. Technol. Lett.*, 5, 787–790, 1993.

58. A.D. Ellis and D.M. Spirit, "Compact 40 Gbit/s optical demultiplexer using a GaInAsP optical amplifier," *Electron. Lett.*, 29, 2115–2116, 1993.

59. K. Suzuki, K. Iwatsuki, S. Nishi, and M. Saruwatari, "Error-free demultiplexing of 160 Gbit/s pulse signal using optical loop mirror including semiconductor laser amplifier," *Electron. Lett.*, 30, 1501–1503, 1994.

60. S. Nakamura, Y. Ueno, and K. Tajima, "168 Gb/s all-optical wavelength conversion with a symmetric-Mach-Zehnder-type switch," *IEEE Photon. Technol. Lett.*, 13, 1091–1093, 2001.

61. S. Nakamura, Y. Ueno, and K. Tajima, "Error-free all-optical demultiplexing at 336 Gb/s with a hybrid-integrated Symmetric Mach-Zehnder all-optical switch," OFC' 02, FD3(PD).

62. Y. Shibata, Y.Tohmori, T. Oku, and T. Itoh, "Semiconductor monolithic wavelength converter: SIPAS," *NTT R&D*, 51, 706–712, 2002. In Japanese.

63. J. Li, B. Olsson, M. Karlsson, and P. Andrekson, "OTDM demultiplexer based on XPM-induced wavelength shifting in highly nonlinear fiber," OFC 2003, TuH6.

64. S. Kodama, T. Yoshimatsu, and H. Ito, "320 Gbit/s error-free demultiplexing operation of a monolithic PD-EAM optical gate," OFC 2003, ThX5.

65   K. Kikuchi, Y. Fukuchi, A. Suzuki, D. Kunimatsu, and H. Ito, "Ultrafast operation of optical time-division demultiplexer using quasi-phase matched LiNbO device," ECOC 2002, No. 8.4.2.

66. A.D. Ellis, T. Widdowson, X. Shan, G.E. Wickens, and D.M. Spirit, "Transmission of a true single polarisation 40 Gbit/s soliton data signal over 205 km using a stabilised erbium fibre ring laser and 40 GHz electronic timing recovery," *Electron. Lett.*, 29, 990–992, 1993.

67. M. Jinno and T. Matsumoto, "Optical tank circuits used for all-optical timing recovery," IEEE J. Quantum Electron. 28, 895–900, 1992.

68. H. Miyamoto, H. Kawakami, T. Kataoka, and K. Hagimoto, "New optical frequency comb gain spectrum generated by stimulated Brillouin amplifier," *Techn. Digest of Optical Amplifiers and their Applications*, OSA Techn. Digest Series 14, Opt. Soc. America, Washington, DC, 1993, p. 194

69. M. Jinno and T. Matsumoto, "All-optical timing extraction using a 1.5 μm self pulsating multielectrode DFB LD," *Electron. Lett.*, 24, 1426–1427, 1988.

70. P.E. Barnsley, G.E. Wickens, H.J. Wickes, and D.M. Spirit, "A 4 × 5 Gb/s transmission system with all-optical clock recovery," *IEEE Photon. Technol. Lett.*, 4, 83–86, 1992.

71. D.J. As, R. Eggemann, U. Feiste, M. Möhrle, E. Patzak, and K. Weich, "Clock recovery based on a new type of self-pulsation in a 1.5 μm two-section InGaAsP/InP DFB laser," *Electron. Lett.*, 29, 141–142, 1993.

72. B. Sartorius, C. Bornholdt, O. Brox, H.J. Ehrke, D. Hoffmann, R. Ludwig, and M. Möhrle, "All-optical clock recovery module based on self-pulsating DFB laser," *Electron. Lett.*, 34, 1664–1665, 1998.

73. C. Bornholdt, B. Sartorius, and M. Möhrle, "All-optical clock recovery at 40 Gbit/s," *Proc. 25th Europ. Conf. Opt. Commun. (ECOC '99)*, Nice, France, 1999, postdeadline papers, p. 54.

74. C. Bornholdt, J. Slovak, M. Möhrle, and B. Sartorius, "Application of a 80 GHz all-optical clock in a 160 km transmission experiment," OFC 2002, TuN6.

75. K. Takayama, K. Habara, and A. Himeno, "High-frequency operation of an all-optical synchronization circuit," *Springer Series in Elect. and Photon.*, Vol.29, Photonic Switching II, Springer, Berlin, 1990, pp. 374–377.

76. T. Ono, Y. Yamabayashi, and Y. Sato, "5 Gbit/s all optical clock recovery circuit using dual mode locking technique," *Conf. Opt. Fiber Commun. (OFC'94)*, Vol. 4, OSA Techn. Digest Series, Opt. Soc. America, Washington, DC, 1994, p. 233.

77. T. Ono, T. Shimizu, Y. Yana, and H. Yokoyama, "Optical clock extraction from 10 Gb/s data pulses using monolithic mode-locked laser diode," *Proc. IEICE Japan National Conference*, Mar.1995, paper No.B-1157. In Japanese.

78. T. Miyazaki, H. Sotobayashi, and W. Chujo, "Optical sampling and DEMUX of 80 Gb/s OTDM signals with optically recovered clock by injection mode-locked laser diode," ECOC'02, No. 2.1.6.

79. T. Ohno, K. Sato, T. Shimizu, T. Furuta, and H. Ito, "40-GHz optical clock recovery from a 160 Gbit/s optical data stream using a regeneratively mode-locked semiconductor laser," ECOC'02, No. 8.4.4.

80. K. Smith and J.K. Lucek, "All-optical clock recovery using a mode-locked laser," *Electron. Lett.*, 28, 1814–1816, 1992.

81. A.D. Ellis, K. Smith, and D.M. Patrick, "All optical clock recovery at bit rates up to 40 Gbit/s," *Electron. Lett.*, 29, 1323–1324, 1993.

82. M. Patrick, "Modelocked ring laser using nonlinearity in a semiconductor laser amplifier," *Electron. Lett.*, 30, 43–44, 1994.

83. K. Takayama, K. Habara, H. Miyazawa, and A. Himeno, "High-frequency operation of a phase-locked loop-type clock regenerator using a 1 × 2 optical switch as a phase comparator," *IEEE Photon. Technol. Lett.*, 4, 99–101, 1992.

84. S. Kawanishi and M. Saruwatari, "New-type phase-locked loop using travelling-wave laser-diode optical amplifier for very high-speed optical transmission," *Electron. Lett.*, 24, 1452–1453, 1988.

85. S. Kawanishi, H. Takara, M. Saruwatari, and T. Kitoh, "Ultrahigh-speed clock recovery circuit using a traveling-wave laser diode amplifier as a 50 GHz phase detector," Topical Meeting on Optical Amplifiers and Their Applications '93, Yokohama, Japan, 1993, postdeadline paper PD5.

86. O. Kamatani, S. Kawanishi, and M. Saruwatari, "Prescaled 6.3 GHz clock recovery from 50 Gbit/s TDM optical signal with 50 GHz PLL using four-wave mixing in a traveling-wave laser diode optical amplifier," *Electron. Lett.*, 30, 807–809, 1994.

87. S. Kawanishi, T. Morioka, O. Kamatani, H. Takara, and M. Saruwatari, "Time-division-multiplexed 100Gbit/s, 200 km optical transmission experiment using PLL timing extraction and all-optical demultiplexing based on polarization insensitive four-wave-mixing," *Conf. Opt. Fiber Commun. (OFC'94)*, OSA Techn. Digest Series, Opt. Soc. America, Washington, DC, 1994, postdeadline paper PD23.

88. O. Kamatani and S. Kawanishi, "Prescaled timing extraction from 400 Gb/s optical signal using a phase lock loop based on four-wave-mixing in a laser diode amplifier," *IEEE Photon. Technol. Lett.*, 8, 1094–1096, 1996.

89. T. Saito, Y. Yano, and N. Henmi, "10-GHz timing extraction from a 20-Gbit/s optical data stream by using a newly proposed optical phase-locked loop," *Conf. Opt. Fiber Commun. (OFC' 94)*, OSA Techn. Digest Series, Opt. Soc. America, Washington, DC, 1994, p. 61.

90. T. Yamamoto, C. Schmidt, E. Dietrich, C. Schubert, J. Berger, R. Ludwig, and H. G. Weber, "40GHz optical clock extraction from 160 Gbit/s data signals using PLL-based clock recovery," OFC'02, TuN5.

91. C. Boerner, C. Schubert, C. Schmidt, E. Hilliger, V. Marembert, J. Berger, S. Ferber, E. Dietrich, R. Ludwig, and H. G. Weber, "160 Gbit/s clock recovery with electro-optic PLL using a bidirectionally operated electroabsorption modulator as phase detector," OFC 2003, FF3.

92. T. Ohara, H. Takara, K. Mori, S. Kawanisiki, S. Mino, T. Yamada, M. Ishii, T. Kitoh, T. Kitagawa, K.R. Parameswaran, and M.M. Fejer, "160 Gbit/s timing extraction using PLL with optical phase modulator and periodically-poled lithium niobate," IEICE Spring Annual Meeting, No.B-10-101, 2003. In Japanese.

93. M. Jinno and M. Abe, "All-optical regenerator based on nonlinear fibre Sagnac interferometer," *Electron. Lett.*, 28, 1350–1352, 1992.

94. K. Smith and J.K. Lucek, "All-optical signal regenerator," *Conf. Lasers and Electro-Optics*, (CLEO'93), 1993, postdeadline paper CPD23.

95. H. Takara, S. Kawanishi, T. Morioka, K. Mori, and M. Saruwatari, "100 Gbit/s optical waveform measurement with 0.6 ps resolution optical sampling using subpicosecond supercontinuum pulses," *Electron. Lett.*, 30, 1152–1153, 1994.

96. H. Takara, S. Kawanishi, A. Yokoo, S. Tomaru, T. Kitoh, and M. Saruwatari, "100 Gbit/s optical signal eye-diagram measurement with optical sampling using organic non-linear optical crystal," *Electron. Lett.*, 32, 2256–2258, 1996.

97. Jie Li, J. Hansryd, P.O. Hedekvist, P.A. Andrekson, and S.N. Knudsen, "300-Gb/s eye-diagram measurement by optical sampling using fiber-based parametric amplification," *IEEE PTL.*, 13, 987–989, 2001.

98. C. Schmidt, F. Futami, S. Watanabe, T. Yamamota, C. Schubert, J. Berger, M. Kroh, H.-J. Ehrke, E. Dietrich, C. BÖrner, R. Ludwig, and H.G. Weber, "Complete optical sampling system with broad gap-free spectral range for 160 Gbit/s and 320 Gbit/s and its application in a transmission system," OFC 2002, ThU1.

99. N. Yamada, N. Banjo, H. Ohta, S. Nogiwa, and Y. Yanagisawa, "320 Gbit/s eye diagram measurement by optical sampling system using a passively mode-locked fiber laser," OFC 2002, ThU3.

100. Jie Li, M. Westlund, H. Sunnerud, B.-E. Olsson, M. Karlsson, and P. A. Andrekson, "0.5 Tb/s eye-diagram measurement by optical sampling using XPM-induced wavelength shifting in highly nonlinear fiber," ECOC 2003, Mo4.6.4.

101. N. Yamada, S. Nogiwa, and H. Ohta, "Measuring eye diagram of 640 Gb/s OTDM signal with optical sampling system by using wavelength-tunable soliton pulse," ECOC 2003, Mo 4.6.5.

102. M. Westlund, H. Sunnerud, M. Karlsson, J. Hansryd, J. Li, P.O. Hedekvist, and P.A. Andrekson, "All-optical synchronous Q-measurements for ultrahigh speed transmission systems," OFC 2002, ThU2.

103. I. Shake, H. Takara, and S. Kawanishi, "Simple Q factor monitoring for BER estimation using opened eye diagrams captured by high-speed asynchronous electrical sampling," *IEEE Photon. Tech. Lett.*, 15, 620–622, 2003.

104. M. Saruwatari, "All-optical signal processing for Terabit/second optical transmission," *IEEE J. Select. Top. Quantum Elect.*, 6, 1363–1374, 2000.

Index

249

9 780367 390273